AF595470

Environmental and Energy Efficiency Evaluation Based on Data Envelopment Analysis (DEA)

Environmental and Energy Efficiency Evaluation Based on Data Envelopment Analysis (DEA)

Editors

Ramon Sala-Garrido
María Molinos-Senante

MDPI • Basel • Beijing • Wuhan • Barcelona • Belgrade • Manchester • Tokyo • Cluj • Tianjin

Editors
Ramon Sala-Garrido
Economic and Business
Mathematics Department,
University of Valencia
Spain

María Molinos-Senante
Department of Hydraulic and
Environmental Engineering,
Pontificia Universidad Católica de Chile
Chile

Editorial Office
MDPI
St. Alban-Anlage 66
4052 Basel, Switzerland

This is a reprint of articles from the Special Issue published online in the open access journal *Energies* (ISSN 1996-1073) (available at: https://www.mdpi.com/journal/energies/special_issues/Environmental_Energy_Efficiency_Evaluation_Based_Data_Envelopment_Analysis).

For citation purposes, cite each article independently as indicated on the article page online and as indicated below:

LastName, A.A.; LastName, B.B.; LastName, C.C. Article Title. *Journal Name* **Year**, *Volume Number*, Page Range.

ISBN 978-3-0365-0572-5 (Hbk)
ISBN 978-3-0365-0573-2 (PDF)

Contents

About the Editors

Ramon Sala-Garrido holds a PhD in Economics. He is a full professor at the Economics Faculty at the University of Valencia (Spain).

María Molinos-Senante holds a PhD in Local Development and Territory. She is an associate professor at the Engineering School at the Pontificia Universidad Católica de Chile (Chile).

Preface to "Environmental and Energy Efficiency Evaluation Based on Data Envelopment Analysis (DEA)"

Most countries have increased their energy generation and use based on fossil fuels, whose low prices have facilitated economic development and the well-being of society. However, fossil fuels, largely responsible for climate change, must give way to cleaner energies and thus avoid the technological gap between developed and developing countries.

The measurement of energy efficiency is a key issue to orient the development of these new energy sources that have to propitiate the change in the production and consumption models. Energy efficiency also has to enhance the preservation and repair of the current environment, so that the negative effects that the consumption of fossil fuels have on people's health are also eliminated.

To measure energy efficiency, data envelopment analysis models are relevant tools where inputs, desirable outputs and undesirable outputs are jointly considered, providing a holistic approach. In this context, the articles published in this Special Issue are a first step to develop new investigations that allow us to better understand the impact of climate change on the environment.

Ramon Sala-Garrido, María Molinos-Senante
Editors

Article

Evaluating the Efficiency of Transit-Oriented Development Using Network Slacks-Based Data Envelopment Analysis

Eun Hak Lee [1], Hosuk Shin [1], Shin-Hyung Cho [2], Seung-Young Kho [3] and Dong-Kyu Kim [3,*]

[1] Department of Civil and Environmental Engineering, Seoul National University, Seoul 08826, Korea; eunhak@snu.ac.kr (E.H.L.); hosukshin@snu.ac.kr (H.S.)
[2] Institute of Engineering Research, Seoul National University, Seoul 08826, Korea; shinhyungcho@snu.ac.kr
[3] Department of Civil and Environmental Engineering and Institute of Construction and Environmental Engineering, Seoul National University, Seoul 08826, Korea; sykho@snu.ac.kr
* Correspondence: dongkyukim@snu.ac.kr; Tel.: +82-2-880-7348

Received: 19 June 2019; Accepted: 15 September 2019; Published: 21 September 2019

Abstract: The purpose of this research is to evaluate transit-oriented development (TOD) efficiency in Seoul using the network slacks-based measure data envelopment analysis (NSBM DEA) model. The smartcard data and socio-economic data are used to evaluate the transit efficiency of 352 subway station areas in Seoul. To measure the TOD efficiency, the two-stage network is designed with the transit design stage and the transit efficiency stage. The overall efficiency score of each station area is estimated through each score of the stage. The results of the efficiency evaluation by station area indicate that the overall efficiency score average is 0.349, with the transit design score and efficiency score estimated to be 0.453 and 0.245, respectively. The results indicate that the balance of each stage is crucial to achieve an efficient station in the concept of transit efficiency. With the efficiency scores of the 352 subway station areas, the TOD efficiency is also evaluated by the administrative units in Seoul. The results of district analysis reveal that the top 10 efficient administrative units are characterized by both residential and commercial land use. The results indicate that efficiency is found to be good in areas having both residential and commercial characteristics.

Keywords: transit-oriented development (TOD); transit efficiency; smartcard data; network slacks-based measure data envelopment analysis (NSBM DEA)

1. Introduction

Transit-oriented development (TOD) is a strategy of urban development that maximizes the transit accessibility to urban areas within walking distance [1]. The main purpose of TOD is to increase transit user comfort and alleviate automobile use by creating an accessible public transportation environment within the city [2]. Moving towards a transit-oriented approach, personal mobility can be prevented and looking in the long-term, more sustainable cities can be built [3].

TOD is practiced in densely populated areas with high demand for transit system facilities such as subway and bus stations. When planning a new public transportation system, organizers must first contemplate the socio-economic characteristics of the area and plan transportation facilities accordingly to the population density, land use, commercial facilities, and residences in the area [4]. The travel demand for transit systems increases when stations are newly introduced into an urban area. Early TODs focused on increasing the connectivity between urban planning and public transit to address problems such as urban sprawl, traffic congestion, and environmental degradation [5]. The development of the TOD concept was aimed at urban planning with a focus on public transit within the city [6]. Historically, TOD has been recognized as an efficient development strategy in terms of the transit environment and socio-economic characteristics [7]. Calthorpe [1] stated that research on public transit use via high-density, multi-purpose land-use patterns could help shape a culture toward becoming a transit-friendly environment. This concept considered the regional factors, such as population density, complex land use, trip purpose, trip frequency, trip demand, and mode [2]. Also, various studies established the TOD evaluation criteria to achieve a more efficient and sustainable transit environment and solve complex urban problems. Most previous research stated that efficient and sustainable transit must balance the social and economic aspects of the transportation environment [8–10]. It means that land use, population density, and residential environment should be closely examined in urban planning [9]. To maximize efficiency at the economic level, transit capacity must be met and maintained to minimize the consumption of resources [11]. The efficiency of TOD also has been explored in previous studies [12–18]. They established one or more indicator variables to quantify the efficiency of TOD strategies. Renne and Wells [16] identified various useful TOD indicators from monitoring successive TODs. Galelo et al. [17] found that travel volume was one of the most representative indicators for evaluating TOD efficiency. However, Yu et al. [18] suggested that a single indicator had a limitation to effectively measure the performance of TOD efficiency and multi-indicators were required.

The massive data created by the Internet of Things (IoT) in cities now enables data scientists to analyze the objective functions of TOD. With its data-driven approaches, data envelopment analysis (DEA) has been widely used to measure the effectiveness of the operation or management of transportation systems [19]. DEA has a significant advantage compared to parametric models as it does not require weight parameters to measure efficiency [20]. The merit of DEA is its simplicity in estimating efficiency with multiple inputs and outputs [19]. DEA also has merit compared to the parametric model. The parametric model assumes a specific production function for the relationship between input and output [18]. However, DEA does not make assumptions about the production function and the given data are utilized to estimate the production relationship between input and output [21]. Therefore, it is possible to avoid the error of setting the type of distribution according to the arbitrary judgment of the analyst. It is possible for the network DEA model to be designed in the order of stages which is required for the evaluation process [22,23]. It is also used with the slacks-based measure (SBM) model for direct comparisons between the observations [24]. The early DEA models have the disadvantage that the inefficiency cannot be directly compared between different observations [25], but the SBM model allows direct comparison between different observations by measuring efficiency based on the slack ratio [26].

The evaluation of the efficiency of the transit system has been performed in previous studies using DEA model [27–31]. With the introduction of automatic fare collection (AFC) system, the data-driven approach for the transit-related analysis has become possible [32]. Various kinds of data, i.e., smartcard data, socio-economic data, and geographical data, were combined to evaluate the efficiency of TOD [33]. Transit efficiency was defined through the relationship between multiple inputs and outputs [34]. Transit efficiency refers to how well the transit system was introduced and managed with respect to the socio-economics, transit infrastructures, and transit trips of each station area [8]. Regarding the TOD concept, both transit design and efficiency must be considered to determine transit efficiency [16,17].

The purpose of this research is to evaluate transit efficiency in Seoul using the network slacks-based measure (NSBM) DEA model. The smartcard data and socio-economic data were used to evaluate the transit efficiency of 352 subway station areas in Seoul. The evaluation process is developed as a two-stage network with transit design and efficiency stage. In the evaluation process, the two-stage NSBM DEA model was used to measure efficiency. The two-stage network constructed with transit design stage and transit efficiency stage. With the results gathered from each stage, the overall efficiency was measured to evaluate the transit efficiencies of the subway station areas. Each station evaluated were grouped by each Seoul administrative unit and ranked based on its TOD efficiency.

2. Methodology for Evaluating Transit Efficiency

2.1. Concept of Data Envelopment Analysis (DEA)

DEA is a nonparametric method for estimating production frontiers. The DEA model identifies relative efficiencies using a number of input and output variables [20,21]. The purpose of measuring efficiency using the DEA model is to determine the strategy of an enterprise, organization, or industry. The first DEA model was developed to evaluate the efficiency and increase production for farm yield in the UK [25]. With respect to productivity, DEA has been applied to linear programming and has been used in various fields [35]. The relative efficiencies of decision-making units (DMUs) were determined and their performances compared. The original model was known as the Charnes, Cooper & Rhodes (CCR) model, which was employed to achieve constant returns to scale (CRS). Since the CRS condition assumes that the unit of production is kept constant at the optimal scale, the input and output are scaled proportionally. The CCR model is the most important model as it shows the most abbreviated methodological features. The CCR model estimates a ratio that can reduce the input as much as possible while keeping the output constant, and vice-versa. For example, there are some considerations to estimate the efficiency score with the input-oriented CCR model. The efficiency score is estimated by summing the weights of the output variables. The summed weights of output variables are between 0.0 and 1.0 score. With the observed J stations ($j = 1, \ldots, J$), each station produces the M outputs using N inputs. The ratio of the input value versus output value is the efficiency score θ and the objective function is to minimize the θ^i which is the reduced ratio of the input variables of target station i. The input-oriented CCR model, therefore, measures the weights of input and output variables to minimize the θ^i, and the efficiency score is estimated by weights of variables. The maximum value of the efficiency score is equal to or less than 1.0 value with the constraints, i.e., $y, x > 0$ and $\lambda \geq 0$. The efficient stations consist of the production frontier, and the inefficient stations improve efficiency in the near direction of the production frontier. In other words, the efficient stations are the reference group that the inefficient station benchmarks to improve its efficiency. The mathematical expression of the input-oriented CCR model is as follows:

$$\theta^{i^*} = Min\ \{\theta^i - \varepsilon\left(\sum_{m}^{M} s_m^- + \sum_{n}^{N} s_n^+\right)\} \tag{1}$$

subject to:

$$\begin{aligned} \theta^i x_m^i &= \sum_{j=1}^{J} x_m^j \lambda^j + s_m^- \\ y_r^i &= \sum_{j=1}^{J} y_r^j \lambda^j - s_r^+ \\ \lambda^j \geq 0,\ & s_m^- \geq 0,\ s_r^+ \geq 0 \end{aligned}$$

where θ^{i^*} is the efficiency score of the target station area i, j ($j = 1, \ldots, J$) is the number of observed station areas, y_n^j is the output number of each variable r ($r = 1, \ldots, R$) of a station area j, x_m^j is the input variable m ($m = 1, \ldots, M$) of a station area j, y_r^j is the output variable r of a station area j, λ^j is the intensity vector of station area j, s_m^- is the slack vector of the input variable x_m, s_o^{k+} is the slack vector of the output variable y_r.

2.2. Network Slacks-Based Measure DEA

The NSBM DEA is used to measure the efficiency of the network that consists of two or more stages. In general, measuring the efficiency with DEA model involves two stages, an input stage, and an output stage. However, network DEA has more than two stages that include intermediate processes [22,23]. These intermediate processes are linking activities that occur in stages of production or that occur internally in DMUs. In other words, the output results of the first stage can be applied to the second stage as inputs for the final result [22]. If the network becomes complex, early DEA models had a limitation with only contain one stage while the network DEA solves this problem by having multiple stages [24]. Since the transit system also has a complex network, the network DEA is suitable for measuring transit efficiency.

The transit infrastructure relative to the socio-economic is important in terms of the transit design [36]. Transit efficiency is defined as the number of transit trips relative to the infrastructures [37]. In this research, the transit design and transit efficiency stages were designed to measure transit efficiency. The process of TOD proceeds with three factors, i.e., socio-economic, transit infrastructure, and transit trip [11]. First, the socio-economic factors are examined to find the area where the transit is needed. Second, the transit infrastructures are built to the area selected from the socio-economic factors. Finally, the transit trips are derived through the socio-economic and transit infrastructures. These three factors are grouped by two stages which are transit design stage and transit efficiency stage. Figure 1 shows the framework of the process used to evaluate transit efficiency. The first stage, the transit design stage was measured by comparing the transit infrastructures with socio-economic factors. For the second stage, the transit efficiency was estimated by comparing the transit trips with transit infrastructures. The overall efficiency score is obtained by the average sum or weighted multiplication of each stage output from the design and efficiency stage. The weight of each stage can be determined by the purpose of the research or the characteristics of the subject [38]. Regarding the TOD evaluation, a bunch of previous studies considered the transit-related factors at the same level [11–13]. Since the concept of TOD considers design and efficiency at the same level, the importance of both stages is considered equal. In this research, identical weights are given at both stages in this research, i.e., w^1 : 0.5 and w^2 : 0.5. In other words, the assumption is made that both the transit design and efficiency stages are equal contributors to the overall efficiency score. With this assumption, the comprehensive efficiency of the station area could be obtained with consideration of the transit design and efficiency.

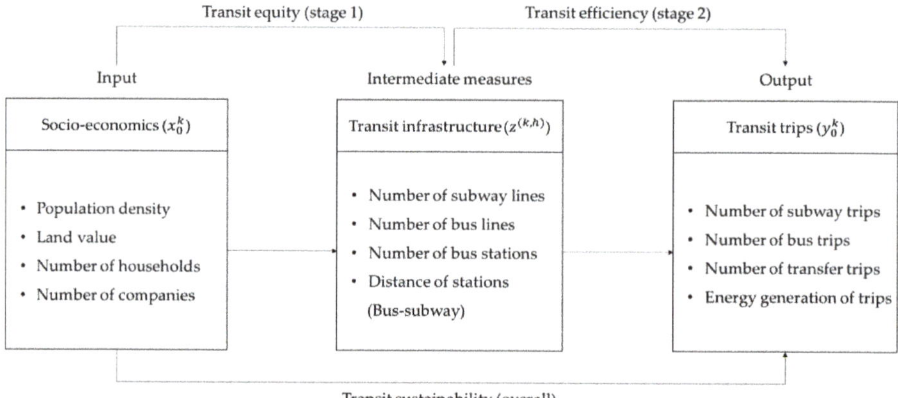

Figure 1. Network framework for measuring transit efficiency.

The production possibility set of network DEA is denoted as $P_{network}$ and its mathematical expression is shown in Equations (2)–(7). The term $z^{(k,h)}$ is an intermediate measure for evaluating transit efficiency, and Equation (5) show their mathematical expressions. The $z^{(k,h)}$ is applied to $z^{(1,2)}$ in this study, since the network framework for measuring the TOD efficiency requires a connection link from the transit design stage to the transit efficiency stage. Equation (5) describes an intermediate measure such as the weights of outputs from the design stage and inputs for the efficiency stage. Equation (6) is used to determine the variable returns to scale (VRS) condition. In the absence of Equation (6), CRS is assumed:

$$P_{network} = (x^k,\ y^k, z^{(k,h)}) \tag{2}$$

$$x^k \geq \sum_{j=1}^{J} x_j^k \lambda_j^k, \forall k \quad (3)$$

$$y^k \leq \sum_{j=1}^{J} y_n^k \lambda_n^k, \forall n, k \quad (4)$$

$$z^{(k,h)} = \begin{cases} \sum_{j=1}^{J} z_j^{(k,h)} \lambda_j^k, \ \forall k, h \text{ (as outputs from } k) \\ \sum_{j=1}^{J} z_j^{(k,h)} \lambda_j^h, \ \forall k, h \text{ (as inputs to } h) \end{cases} \quad (5)$$

$$\sum_{j=1}^{J} \lambda_j^k = 1, \ \forall k \quad (6)$$

$$\lambda_j^k \geq 0, \ \forall j, k \quad (7)$$

where $P_{network}$ is the production possibility set, k ($k = 1,\ 2$) is the number of stages that $k = 1$ is the transit design stage and $k = 2$ is the transit efficiency stage, j ($j = 1, \ldots, J$) is the number of observed station areas, $x_{mj}^k \in R_+^{m_k}$ is the input variable of station area j of stage k, $y_{rj}^k \in R_+^{r_k}$ is the output variable of station area j of stage k, $(k,\ h) \in L$ is the connection link from transit design stage to transit efficiency stage, $z^{(k,h)} \in R_+^{(k,h)}$ is an intermediate measure from the transit design stage to transit efficiency stage, λ^k is the intensity vector corresponding to stage k, $z_j^{(k,h)} \lambda_j^k$ is the outputs from the transit design stage, and $z_j^{(k,h)} \lambda_j^h$ is the inputs to the transit efficiency stage.

The SBM DEA is a widely used model for evaluating efficiency [23]. Network DEA with a slacks-based approach was first developed by Tone and Tsutsui [22], and this model is called NSBM DEA model. The NSBM DEA model evaluates efficiency using the input and output slack. The slack is the difference value amount from the desired value amount from the actual input and output variables [24]. The two slack values are estimated irrespective of the variable unit and are calculated as an efficiency measure using the average of the reduced inputs and the average of the increased outputs. Since input and output variables have different units, slack values are converted to ratio values by dividing the original observation values. NSBM DEA is performed by calculating the slack between observations and production changes. NSBM DEA is referred to as a SBM because it is calculated based on the slack between the observations and production changes. The most important feature of NSBM DEA is that the measure of efficiency does not change even when the units of the input or output variables change. This is because the input or output slack is calculated as a ratio and is thus independent of the unit. Compared to the early DEA models, SBM model has the advantage of allowing direct comparison between different DMUs [23]. Since the efficiency score of early DEA model is estimated by adding the slacks of variables with different units, the inefficiency cannot be directly compared between different DMUs [28]. The efficiency of SBM model is measured by adding the slack ratio. Since SBM model uses the slack ratio of each variable, the efficiency can be measured regardless of the units of the variables [24].

NSBM DEA has three variations, i.e., the non-, input-, and output-oriented models. These three models can be employed depending on the objective or the features of the variables. In this research, we used the output-oriented SBM, for which the output direction is improving efficiency. The output-oriented SBM measures efficiency by fixed inputs and maximizing outputs. Since the outputs of each stage of transit analysis are required by the given conditions, the use of the output-oriented SBM is reasonable for determining the transit design and efficiency scores. It is difficult to change land use or eliminate existing facilities. Therefore, it is necessary to derive efficiency rankings by maximizing the output variables of each stage. NSBM DEA is widely used for evaluating relative efficiencies because it measures the efficiencies of DMUs. Given the transit system features mentioned above, the NSBM DEA model is suitable for evaluating transit efficiency. The mathematical expression for measuring transit efficiency is shown in Equation (8).

$$\theta_i^* = min\frac{\sum_{k=1}^{K} w^k[1 - \frac{1}{m_k}(\sum_{m=1}^{m_k} s_{mi}^{k-}/x_{mi}^k)]}{\sum_{k=1}^{K} w^k[1 + \frac{1}{r_k}(\sum_{r=1}^{r_k} s_{ri}^{k-}/y_{ri}^k)]} \quad (8)$$

subject to:

$$\begin{aligned}
&\sum_{k=1}^{K} w^k = 1,\ \forall k\\
&w^k \geq 0,\ \forall k\\
x_{mi}^k &= \sum_{j=1}^{J} x_{mj}^k \lambda_j^k + s_m^{k-},\ \forall m, k\\
y_{ri}^k &= \sum_{j=1}^{J} y_{rj}^k \lambda_j^k - s_r^{k+},\ \forall r, k\\
&\sum_{j=1}^{J} \lambda_j^k = 1, \forall k\\
&\lambda_j^k \geq 0,\ \forall j, k\\
&s_m^{k-} \geq 0,\ \forall m, k\\
&s_r^{k+} \geq 0, \forall r, k
\end{aligned}$$

where θ_i^* is the overall efficiency score of station area i, k (k = 1, 2) is the number of the stages that $k = 1$ is the transit design stage and $k = 2$ is the efficiency stage, w^k is the relative weight of stage k, x_{mi}^k is the input variable m of station area i of stage k, y_{ri}^k is the output variable r of the station area i of stage k, λ^k is the intensity vector corresponding to the stage k, s_m^{k-} is the slack vector of the input variable x_m of the stage k, and s_r^{k+} is the slack vector of the output variable y_r of the stage k.

3. Data Description

3.1. Description of Smartcard Data

The government of Seoul has been operating the AFC system such as the smartcard system since 2004. The transit fare from the origin to destination is charged based on the total distance traveled by buses, subways, or both. Within the AFC system, travelers can use any combination of transit modes [31]. Since the transit system in Seoul has been operating as a 100% smartcard system, it is possible to extract 99% of the transit trip information. The smartcard data in Seoul records about 20 million individual transit trips per day. Each individual item of information is classified with respect to 36 categories, including card ID, boarding station, alighting station, boarding time, alighting time, etc. With the smartcard data, it is possible to obtain the numbers of subway, bus, bus–bus transfer, and subway–bus transfer trips. Since the smartcard data in Seoul include all transit users' trips, it can be used to analyze transit efficiency by station area [27]. Among the 36 categories of smartcard data, 11 were used in this research. These 11 categories include the card ID, boarding station ID, alighting station ID, number of transfers, alighting time, alighting date, boarding time, boarding date, line ID, vehicle ID, and zone code. Table 1 shows the 36 categories of smartcard data in Seoul of which 11 were used in this research.

Table 1. Description of the smart card data.

No.	Categories	No.	Categories
1	Card ID *	19	Alighting time *
2	Name of the transit line	20	Transaction ID
3	Vehicle number	21	Company name
4	Boarding station ID *	22	Ending run time
5	Alighting station ID *	23	Alighting date *
6	The number of users	24	User division
7	Alighting violation penalty	25	Alighting fare
8	General user code	26	Total travel time
9	Time code	27	Boarding time *
10	Year	28	Boarding date *
11	Mode code	29	Line ID *
12	Company code	30	Vehicle ID *
13	Starting run time	31	Child user code
14	Name of boarding station	32	Name of alighting station
15	Number of transfer *	33	User group
16	Boarding fare	34	Boarding violation penalty
17	Total travel distance	35	Zone code *
18	Student user code	36	Other user code

* indicates the categories used in this research.

3.2. Description of Socio-Economic Data

The government of Seoul has been operating an open big-data portal since October 2013. This open big-data portal is an integrated data platform and provides the public with data about Seoul. Open big-data portal is available to anyone who has the desire to use it. The portal refers to all the data and information produced by public institutions such as public information. It facilitates communication and cooperation between all those interested. These open data are wide-ranging in scope, with information ranging from weather, geography, transportation, and food to historical documents and records. Open data related to socio-economic and environmental indicators are provided at the administrative and statistical aggregation district level. Seoul consists of 424 neighborhood areas, called dong, and each neighborhood is composed of one or more census areas. In this research, the census area unit is used to aggregate the population density, land value, number of households, and number of companies. There are 103,455 census area units in Seoul with an average area of 0.58 km^2.

3.3. Data Preprocessing

The station areas of the 352 subway stations in Seoul were designated as DMUs in order to measure transit efficiency. The Enforcement Rules of the Urban Planning Ordinance of Seoul defines the station area as "an area within a 500 m radius from the center of stations such as subway, national railway, and light rail". [39]. This standard was employed in several previous research related to the transit in Seoul [40,41]. Data preprocessing involved compiling these data by the station area. Transit and socio-economic data were obtained from both the smartcard and open data, respectively.

Socio-economic data from the open data portal were also compiled by the station area. Among the various open data, the population density, land value, number of households, and the number of companies were obtained for this research. Since all the obtained data were provided within census area units, it was necessary to aggregate the data values by the station area. The population density and land use values were compiled by averaging. The number of households and companies were aggregated by summing. From the result of preprocessing, the socio-economic data, population density, land value, the number of households, and the number of companies were determined to average: 34,249 (person/km^2), 5903 (1000 won/m^2), 5249 households, and 134 companies, respectively. Table 2 lists the descriptive statistics of the socio-economic data from the open-data portal.

Table 2. Descriptive statistics of the socio-economic data.

Socio-Economic Data (Input Variables)	Mean	Max.	Min.	Standard Deviation (S.D.)
Population density (person/km^2)	34,249	67,074	342	12,186
Land value (1000 won/m^2)	5903	29,275	79	4133
Number of households	5249	14,914	74	2648
Number of companies	134	198	24	23

The transit data consist of transit infrastructures and trips per transit stations. The infrastructure variables include the numbers of subway lines, bus lines, and bus stations, and the average distance between a bus stop and subway station. The transit trips data include the numbers of subway trips, bus trips, and transfer trips between subway and bus, and the energy consumption by transit trips.

Since there is an overlapping area between some station areas, the average distance between the bus and the subway stations and the number of transfer trips are included as a variable. The definition of energy consumption is the consumed energy by transit mode per trip [42]. The station area the transit energy consumption of the individual station area can be calculated. Since the transit modes consist of subway and bus, the energy consumption is obtained by the sum of each mode's trips multiplied by the conversion factor. For the transfer trips, conversion factors of each mode are multiplied by each trip. The energy consumption by each station area was calculated using conversion factors, i.e., 0.7 for a subway trip (Mcal/trip), and 3.2 for a bus trip (Mcal/trip). These factors are provided by the Ministry of Trade, Industry and Energy of the Republic of Korea [42]. Figure 2 shows the heat-map of transfer trips on station area.

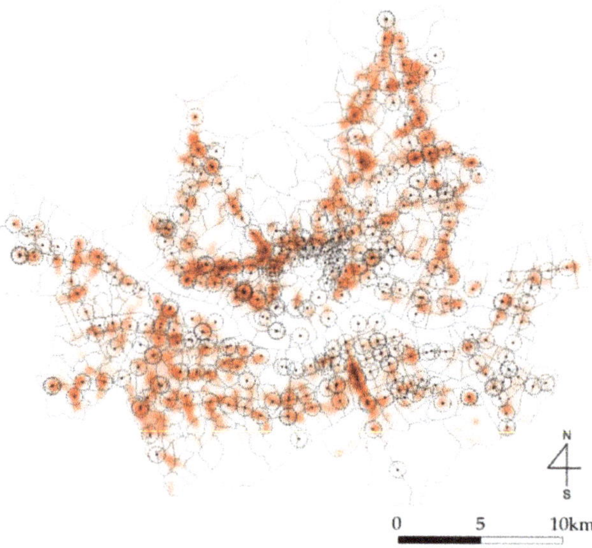

Figure 2. Heat-map of transfer trips on station area.

The data preprocessing results by station area for the numbers of subway lines, bus lines, and bus stations, and the distance between bus stops and subway stations yielded averages of 1.6 lines, 34 lines, 70 stations and 254 m, respectively. To identify the relationship between transit modes, the numbers of bus lines and stations were counted by types of buses, i.e., main bus, branch bus, or circulation bus. The numbers of subway trips, bus trips, and transfer trips, and energy consumption were 36,640 trips, 96,239 trips, 6164 trips, and 377,910 Mcal/trip, respectively. Table 3 lists the descriptive statistics of the transit data obtained from smartcard data.

Table 3. Descriptive statistics of the transit efficiency data.

	Details	Mean	Max.	Min.	S.D.
Transit infrastructures (intermediate variables)	Number of subway lines	1.6	5	1	0.8
	Number of bus lines	34	145	4	22.1
	Number of bus stations	70	321	13	44
	Distance of bus and subway stations (m)	254	433	132	59
Transit trips (output variables)	Number of subway trips	36,640	225,130	1860	30,805
	Number of bus trips	96,239	473,770	2090	85,324
	Number of transfer trips	6164	62,183	26	8171
	Energy consumption (Mcal/trip)	377,910	1,603,120	24,697	268,318

4. Application

4.1. Results of Transit Efficiency

With the results of the NSBM DEA of this research, the overall efficiency score was obtained by the average sum or weighted multiplication of each stage output from the design and efficiency stages. Table 4 and Figure 3 show the efficiency evaluation results for the station areas in Seoul. Based on the efficiency evaluation results in Table 4, the average overall efficiency score is 0.349. The transit design and efficiency evaluation results average 0.453 and 0.245, respectively. Since the overall efficiency score is calculated using the transit design and transit efficiency scores, multiplying by a weight of 0.5 implies that the overall efficiency score is affected more by the design score rather than the efficiency score. Regarding the 0.349 score and 0.207 standard deviation (S.D.), there is clearly a large gap between the efficient and inefficient station areas in Seoul. The 10 station areas were determined to be efficient, i.e., Euljiro 1ga, Shindorim, Gupabal, Dongjak, Yeongdeungpo, Digital media city, Gasan digital, Magok, Bokjung, and Gaehwa station. The top 10 efficient stations are the DMUs with the highest scores of 1.000 in both the transit design and efficiency stages is served as benchmarks for the other 342 inefficient stations. As the means of the input variables for the top 10 efficient station areas, the population density, land value, number of households, and number of companies are 21,581 (person/ km^2), 4478 (1000 won/m^2), 2450 households, and 110 companies, respectively. The means of the intermediate variables for the top 10 efficient station areas are 1.8 subway lines, 55 bus lines, 91 stations, and 212 m distance between stations, respectively. The means of the output variables for the efficient station areas, i.e., the number of subway trips, bus trips, transfer trips, and energy consumption, are 182,420, 97,491, 11,747 trips and 439,666 Mcal/trip, respectively.

Table 4. Efficiency evaluation results for station areas in Seoul.

	Details	Total Stations (352 Stations)		Efficient Stations (Top 10 Stations)		Inefficient Stations (Bottom 10 Stations)	
		Mean	S.D.	Mean	S.D.	Mean	S.D.
Measured score	Overall efficiency score	0.349	0.207	1.000	0.000	0.090	0.011
	Design (stage 1, β_1 : 0.5)	0.453	0.209	1.000	0.000	0.157	0.016
	Efficiency (stage 2, β_2 : 0.5)	0.245	0.251	1.000	0.000	0.023	0.016
Input variable	Population density (person/km^2)	34,249	12,186	21,581	15,853	38,853	11,657
	Land coat (1000 won/m^2)	5903	4133	4478	6097	7,300	4140
	Number of households	5249	2648	2450	2133	5,329	1844
	Number of companies	134	23	110	40	140	13
Intermediate variable (output from stage 1)	Number of subway lines	2.7	0.7	1.8	0.8	3.5	1.1
	Number of bus lines	52	21	55	33	93	36
	Number of bus stations	167	81	91	52	242	38
	Distance of bus and subway stations (m)	231	46	212	64	202	36
Output variable	Number of subway trips	36,640	30,805	182,420	205,228	37,676	58,562
	Number of bus trips	96,239	85,324	97,491	80,063	57,391	29,638
	Number of transfer trips	6164	8171	11,747	11,575	511	620
	Energy consumption (Mcal/trip)	377,910	268,318	439,666	262,293	210,026	107,734

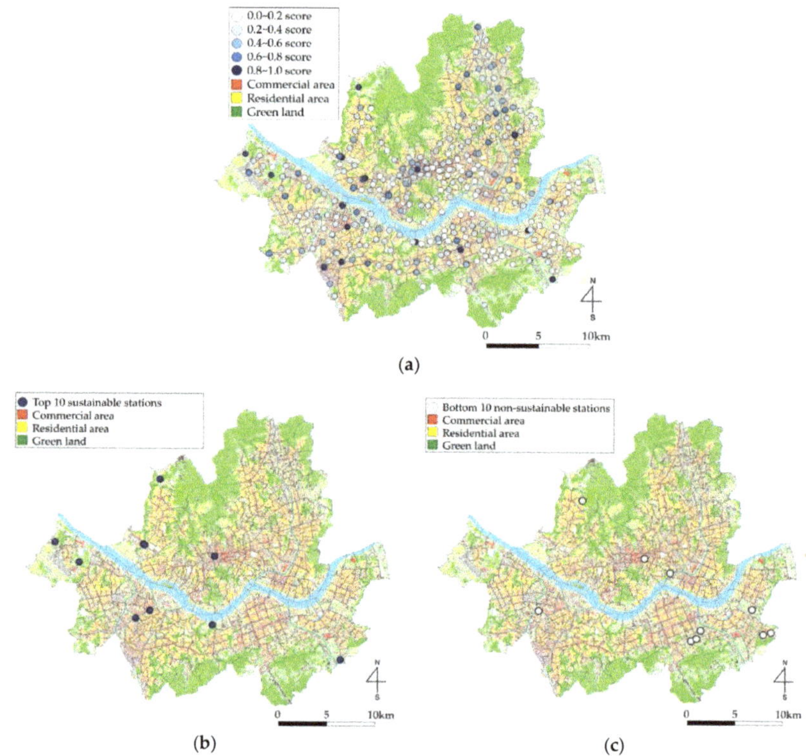

Figure 3. Visualization of the efficiency evaluation results for Seoul: (**a**) efficiency result of the 352 stations; (**b**) top 10 efficient stations; and (**c**) bottom 10 inefficient stations.

A comparison of the efficient station area scores for the 352 station areas shows that all the input variable scores of the efficient station areas are lower than the mean score of all the station areas. In particular, the population density and number of households in the efficient station areas are about 37% and 16% lower than the average value of all the station areas, respectively. From the statistics for the intermediate variables, the number of subway lines and bus stations are about 66% and 54% of the means for all the station areas, and the number of bus lines is almost the same as the mean for all the station areas. These efficient station areas have relatively low population densities and small household. Although the population densities and number of households are smaller than the average values for all the station areas, the transit infrastructures are also well constructed. This is the main reason that transit design is estimated to be high. From the statistics of the output variables, the number of subway trips, bus trips, and transfer trips, and the energy consumption are about 4.98, 1.01, 1.91, and 1.16 times higher, respectively, than the mean values for all the station areas. The transit efficiency score was estimated to be 1.000, since it has a small population density, number of households, and the number of transit infrastructures compared to the output variables. This is because the inputs are lower and outputs are higher than the means for all the station areas, respectively. Since both the design and efficiency scores were 1.000, the overall efficiency score was also estimated to be 1.000.

The balance of each stage is crucial to achieve an efficient station in the concept of transit efficiency. For instance, if the transit infrastructure is well established with low transit trips, the design stage scores will be high but the efficiency stage scores will be relatively low. On the contrary, if the transit infrastructure is not well established with high transit trips, the design stage scores will be low and the efficiency stage scores will be relatively high. Hence, the overall efficient stations have balance in both design and efficiency stage scores. From the results of the evaluation, overall efficient stations are usually located in areas that have a relatively low population density but have well-built transit infrastructures with high transit trips.

The bottom 10 inefficient station areas include Donrimcheon, Hakyoeul, Eungbong, Guryong, Gaepodong, Dongdaemoon, Olympic Park, Geoyeo, Macheon, and Dokbawi stations. These bottom 10 inefficient station areas are relatively estimated to be the lowest overall efficiency scores among the 352 station areas. These 10 relatively inefficient station areas are DMUs in which the TOD should take top priority in improvement. The overall average score of the bottom 10 inefficient station areas was 0.090, which indicates that an improvement of 99.1% is required. The averages of design and efficiency scores were 0.157 and 0.023, respectively, which means that about 84.3% of these areas are poorly designed and 98.8% are inefficient. When targeting the efficient station, transit infrastructure-related variables should be improved by 84.3% compared to the socio-economic related variables. The transit trip-related variables should also be increased by 98.8% compared to the infrastructure-related variables. For the input variables of the bottom 10 inefficient station areas, the population density, land price, number of households, and number of companies totaled 38,853 (persons/km^2), 7300 (1000 won/m^2), 5329 households, and 140 companies, respectively. The intermediate variables, which are the outputs of the design stage, i.e., numbers of subways, bus lines, and bus stations and the distance between bus stops and subway stations averaged 1.2 subway lines, 11 bus lines, 20 stations, and 258 m, respectively. For the outputs, the numbers of subway trips, bus trips, and transfer trips, and the energy consumption totaled 37,676, 57,391, and 511 trips, and 210,026 Mcal/trip, respectively.

A comparison of the statistics for the bottom 10 inefficient station areas with all the station areas was conducted. It shows that all the input values of the bottom 10 inefficient station areas are higher than the mean values of all the stations. With respect to the statistics for the intermediate variables, the numbers of subway lines, bus lines, and bus stations are about 1.3, 1.8, and 1.5 times more, respectively. The distance between bus stops and subway stations is about 13% less than the mean distance for all the station areas. However, the output variable values were lower than the mean values of the 352 station areas. Although the number of subway trips was similar to the mean value for all the station areas, the numbers of bus trips and transfer trips and the energy consumption were about 40%, 92%, 44% less, respectively, than the mean values for all the station areas. Since the population and number of households are higher than the mean values for all the station areas, transit infrastructures must be better equipped than the mean required for all 352 station areas. The transit efficiency scores are also very low since the output values do not meet the required number of transit infrastructures.

4.2. Discussion

Figure 4 shows the overall efficiency scores that consider both the design and efficiency stages. As mentioned above, station areas with a score between 0.000 and 0.200 have top priority for TOD improvement due to low design and efficiency scores. Based on the mean score for each stage, i.e., transit design and efficiency analysis, the DMUs were also divided into four groups to indicate the type of improvement needed at these station areas. In Figure 4b, we can see that both the design and efficiency scores of group 1 are lower than the mean value, which mean that station areas in group 1 need to improve both their design and efficiency. The station areas in group 2 need improvements in design more than efficiency. Station areas in group 3 can improve their overall efficiency by improving their efficiency, and stations areas in group 4 qualify as relatively efficient stations. The types of improvement required to achieve overall efficiency can be determined by the scores of each stage or by group analysis.

The analysis of an administrative unit is needed to practically evaluate the TOD priorities. Using the efficiency evaluation results of the station areas, it is also possible to analyze transit efficiency by the administrative units in Seoul. The dong unit is the smallest unit among the administrative units. The overall efficiency scores of the dong unit were obtained by averaging the scores of the relevant station areas.

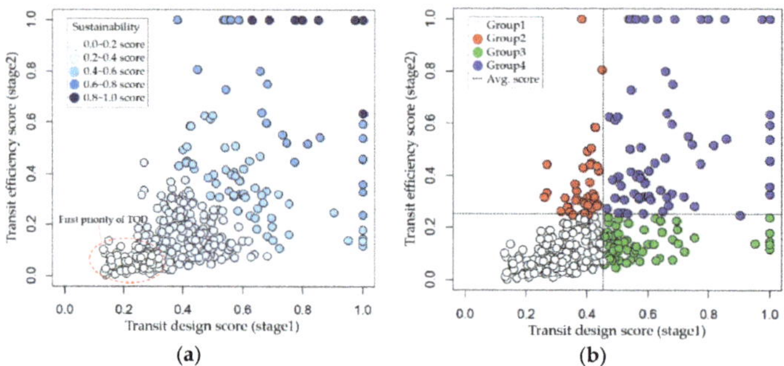

Figure 4. Efficiency evaluation results with respect to design and efficiency: (**a**) efficiency results prioritized regarding TOD; and (**b**) group-wise efficiency results.

The dongs were classified with respect to four dimensions, i.e., non-scoring, low-scoring, mid-scoring, and efficient, and land-use features of each dimension were identified by the socioeconomic variables of each dong. The overall efficiency scores of the dong unit are shown in Table 5 and Figure 5.

Table 5. Results of overall efficiency by Dong unit in Seoul.

Details		Total (424 Dongs)	Non-Scoring (41 Dongs)	Inefficient		Efficient (8 Dongs)
				Low-Scoring (79 Dongs)	Mid-Scoring (296 Dongs)	
Overall efficiency score	Overall efficiency score	0.351	-	0.153	0.387	1.000
	Design (stage 1, β_1 : 0.5)	0.443	-	0.249	0.480	1.000
	Efficiency (stage 2, β_2 : 0.5)	0.259	-	0.058	0.293	1.000
Socio-economics	Population Density (person/km^2)	24,397	22,463	21,960	25,592	14,138
	Land cost (1000 won/m^2)	4454	2410	6208	4290	3682
	Number of households	10,133	9478	9815	10,280	11,232
	Number of companies	1934	1227	2083	2004	1488

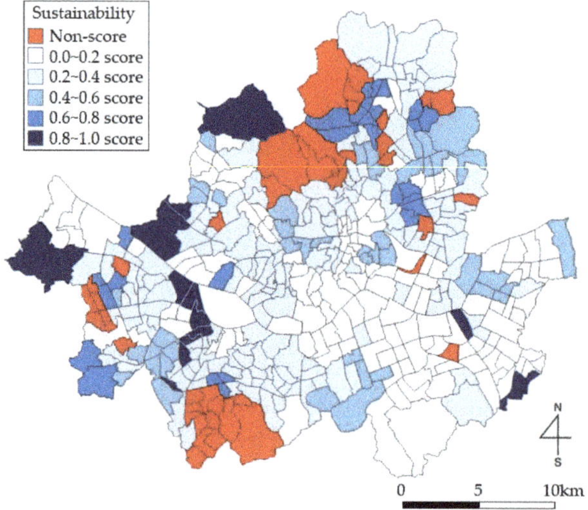

Figure 5. Visualization of the efficiency evaluation results by dong unit in Seoul.

Figure 5 shows the results of the overall transit efficiency of Seoul by the 424 dong administrative units. Since the station areas were set within a 500-m radius, a station area could cover several dongs. Even though the no subway stations reside in the dong, the dong can be affected by the station area radius that spans multiple dongs. The efficiency score of each dong is estimated by calculating the average efficiency values of the station areas that reside or radius spans into the dong.

Eight dongs were estimated to be efficient, having earned a 1.000 overall efficiency score, i.e., Wirye-Dong, Gonghang-Dong, Sangam-Dong, Susek-Dong, Jinkwan-Dong, Daerim (3)-Dong, Dorim-Dong, and Yeongdeungpobon-Dong. These eight dongs can serve as reference DMUs for achieving efficient TOD. Regarding regional characteristics, efficient dongs have both residential and commercial land-use features. The top priority areas for TOD are the red dongs in Figure 5. These 41 dongs were estimated to be non-scoring areas since they contain no station areas. These non-scoring dongs do not have any station areas that reside or spans within the dong despite range being 500 m radius for the station area. Residents of the non-scoring dongs must first take a bus to make transit trips due to the lack of subway stations. Since only bus infrastructures have been established, these areas are at a disadvantage for improving their overall efficiency. With respect to land use characteristics, these top-priority dongs for TOD are mostly residential with a large number of households, i.e., 9478 households. The second-priority areas are the 79 low-scoring dongs, which earned overall efficiency scores in the 0.0 to 0.2 range. Although these 79 dongs were designed for TOD, their transit infrastructures, trips, and energy consumption must be improved to achieve efficient TOD. Regarding land use, these second-priority dongs also have commercial features and host a large number of companies, i.e., 2083 companies.

5. Conclusions

The government of Seoul has been operating the AFC system and open data platform since 2004 and 2013, respectively. These systems provide the opportunity to analyze the efficiency of station areas in terms of TOD. This research was conducted to evaluate the transit efficiency of subway station areas. A total of 352 subway stations within a 500-m radius in Seoul were analyzed. Socioeconomic data were obtained from the open data platform, i.e., population density, land value, number of households, and number of companies. Transit-related data were obtained from Seoul's smartcard data, i.e., the number of subway lines, the number of bus lines, and the number of bus stations. Given the TOD concept, the transit efficiency evaluation was designed as a two-stage network slacks-based measure data envelopment analysis (NSBM DEA). The first stage was designed as a transit-design evaluation and the second stage was evaluated with respect to transit efficiency.

The results of the evaluation were as follows: the overall efficiency score and S.D. were estimated to be 0.349 and 0.207, respectively, which points to a large gap between the efficient and inefficient station areas in Seoul. The analysis results indicated that the eight efficient dongs were characterized by both residential and commercial land use. In addition, overall efficiency was found to be high in areas that have both residential and commercial characteristics. The non-scoring dongs were identified as having top priority for TOD, and the land-use features of these dongs are residential. Dongs with overall efficiency scores in the 0.0 to 0.2 range were designated as second-priority areas, and these dongs also have commercial features. Based on the results of the transit design and efficiency evaluation, it was possible to determine the TOD-related priorities for stations and dongs.

Considering the regional characteristics, development efforts are also required to improve overall efficiency. This research measured transit overall efficiency based on transit design and efficiency. From the results, recommendations regarding the TOD priorities and development directions were made by station areas. Although various factors were used to evaluate TOD efficiency, a variety of additional socioeconomic and transit factors must also be considered. In this study, we defined the station area as the area within a 500-m radius adjacent to a subway station based on previous literature. The overlapping areas may affect the results of the analysis. The score of a station area that contains multiple stations would be different from the score of a station area that only contains one station. Based on the sensitivity analysis, the effect of the station area should be investigated. The weight of each stage of NSBM DEA model can be also changed according to regional conditions, environment, and culture.

Author Contributions: For The authors confirm the contribution to the paper as follows: Study conception and design: E.H.L., H.S., S.-H.C., S.-Y.K., D.-K.K.; Data collection: E.H.L., H.S., S.-H.C., D.-K.K.; Analysis and interpretation of results: E.H.L., H.S., S.-Y.K., D.-K.K.; Draft manuscript preparation: E.H.L., H.S., S.-H.C., D.-K.K. All authors reviewed the results and approved the final version of the manuscript.

Funding: This research was supported by a grant from R & D Program of the Korea Railroad Research Institute, Korea.

Conflicts of Interest: The authors declare no conflict of interest.

References

1. Calthorpe, P. *The Next American Metropolis: Ecology, Community, and the American Dream*; Princeton Architectural Press: New York, NY, USA, 1993.
2. Nasri, A.; Zhang, L. The analysis of transit-oriented development (TOD) in Washington, DC and Baltimore metropolitan areas. *Transp. Policy* **2014**, *32*, 172–179. [CrossRef]
3. Ewing, R.; Cervero, R. Travel and the built environment: A synthesis. *Transp. Res. Rec.* **2001**, *1780*, 87–114. [CrossRef]
4. Papa, E.; Bertolini, L. Accessibility and Transit-Oriented Development in European metropolitan areas. *J. Transp. Geogr.* **2015**, *47*, 70–83. [CrossRef]
5. Knowles, R.D. Transit Oriented Development in Copenhagen, Denmark: From the Finger Plan to Ørestad. *J. Transp. Geogr.* **2012**, *22*, 251–261. [CrossRef]
6. Vale, D.S. Transit-oriented development, integration of land use and transport, and pedestrian accessibility: Combining node-place model with pedestrian shed ratio to evaluate and classify station areas in Lisbon. *J. Transp. Geogr.* **2015**, *45*, 70–80. [CrossRef]
7. Sahu, A. A methodology to modify land uses in a transit oriented development scenario. *J. Environ. Manag.* **2018**, *213*, 467–477. [CrossRef] [PubMed]
8. Li, Z.; Han, Z.; Xin, J.; Luo, X.; Su, S.; Weng, M. Transit oriented development among metro station areas in Shanghai, China: Variations, typology, optimization and implications for land use planning. *Land Use Policy* **2019**, *82*, 269–282. [CrossRef]
9. Zhao, P. Sustainable urban expansion and transportation in a growing megacity: Consequences of urban sprawl for mobility on the urban fringe of Beijing. *Habitat Int.* **2010**, *34*, 236–243. [CrossRef]
10. Wey, W.M.; Zhang, H.; Chang, Y.J. Alternative transit-oriented development evaluation in sustainable built environment planning. *Habitat Int.* **2016**, *55*, 109–123. [CrossRef]
11. Cervero, R.; Kockelman, K. Travel demand and the 3Ds: Density, diversity, and design. *Transp. Res. Part D Transp. Environ.* **1997**, *2*, 199–219. [CrossRef]
12. Cervero, R.; Sarmiento, O.L.; Jacoby, E.; Gomez, L.F.; Neiman, A. Influences of built environments on walking and cycling: Lessons from Bogotá. *Int. J. Sustain. Transp.* **2009**, *3*, 203–226. [CrossRef]
13. Singh, Y.J.; Fard, P.; Zuidgeest, M.; Brussel, M.; Van Maarseveen, M. Measuring transit oriented development: A spatial multi criteria assessment approach for the City Region Arnhem and Nijmegen. *J. Transp. Geogr.* **2014**, *35*, 130–143. [CrossRef]
14. Winstead, D.L. Smart growth, smart transportation: A new program to manage growth in Maryland. *Urb. Law* **1998**, *30*, 537–545.
15. Higgins, C.D.; Kanaroglou, P.S. A latent class method for classifying and evaluating the performance of station area transit-oriented development in the Toronto region. *J. Transp. Geogr.* **2016**, *52*, 61–72. [CrossRef]
16. Renne, J.L.; Wells, J. *Transit-Oriented Development: Developing a Strategy to Measure Success*; Transportation Research Board: Washington, DC, USA, 2005.
17. Galelo, A.; Ribeiro, A.; Martinez, L.M. Measuring and Evaluating the Impacts of TOD Measures–Searching for Evidence of TOD Characteristics in Azambuja Train Line. *Procedia Soc. Behav. Sci.* **2014**, *111*, 899–908. [CrossRef]
18. Yu, X.; Lang, M.; Gao, Y.; Wang, K.; Su, C.H.; Tsai, S.B.; Huo, M.; Yu, X.; Li, S. An Empirical Study on the Design of China High-Speed Rail Express Train Operation Plan—From a Sustainable Transport Perspective. *Sustainability* **2018**, *10*, 2478. [CrossRef]

19. Barnum, D.T.; McNeil, S.; Hart, J. Comparing the efficiency of public transportation subunits using data envelopment analysis. *J. Public Transp.* **2007**, *10*, 1–16. [CrossRef]
20. Banker, R.D.; Charnes, A.; Cooper, W.W. Some models for estimating technical and scale inefficiencies in data envelopment analysis. *Manag. Sci.* **1984**, *30*, 1078–1092. [CrossRef]
21. Charnes, A.; Cooper, W.W.; Rhodes, E. Measuring the efficiency of decision making units. *Eur. J. Oper. Res.* **1978**, *2*, 429–444. [CrossRef]
22. Tone, K.; Tsutsui, M. Network DEA: A slacks-based measure approach. *Eur. J. Oper. Res.* **2009**, *197*, 243–252. [CrossRef]
23. Tone, K. A slacks-based measure of efficiency in data envelopment analysis. *Eur. J. Oper. Res.* **2001**, *130*, 498–509. [CrossRef]
24. Tone, K.; Tsutsui, M. Dynamic DEA: A slacks-based measure approach. *Omega* **2010**, *38*, 145–156. [CrossRef]
25. Farrell, M.J.; Fieldhouse, M. Estimating efficient production functions under increasing returns to scale. *J. R. Stat. Soc. Ser. A* **1962**, *125*, 252–267. [CrossRef]
26. Li, T.; Yang, W.; Zhang, H.; Cao, X. Evaluating the impact of transport investment on the efficiency of regional integrated transport systems in China. *Transp. Policy* **2016**, *45*, 66–76. [CrossRef]
27. Lee, E.H.; Lee, H.; Kho, S.Y.; Kim, D.K. Evaluation of Transfer Efficiency between Bus and Subway based on Data Envelopment Analysis using Smart Card Data. *KSCE J. Civ. Eng.* **2019**, *23*, 788–799. [CrossRef]
28. Hahn, J.S.; Kho, S.Y.; Choi, K.; Kim, D.K. Sustainability evaluation of rapid routes for buses with a network DEA model. *Int. J. Sustain. Transp.* **2017**, *11*, 659–669. [CrossRef]
29. Hahn, J.S.; Kim, D.K.; Kim, H.C.; Lee, C. Efficiency analysis on bus companies in Seoul city using a network DEA model. *KSCE J. Civ. Eng.* **2013**, *17*, 1480–1488. [CrossRef]
30. Hahn, J.S.; Sung, H.M.; Park, M.C.; Kho, S.Y.; Kim, D.K. Empirical evaluation on the efficiency of the trucking industry in Korea. *KSCE J. Civ. Eng.* **2015**, *19*, 1088–1096. [CrossRef]
31. Lee, E.H.; Lee, I.; Cho, S.H.; Kho, S.Y.; Kim, D.K. A Travel Behavior-Based Skip-Stop Strategy Considering Train Choice Behaviors Based on Smartcard Data. *Sustainability* **2019**, *11*, 2791. [CrossRef]
32. Li, X.; Yu, J.; Shaw, J.; Wang, Y. Route-Level Transit Operational-Efficiency Assessment with a Bootstrap Super-Data-Envelopment Analysis Model. *J. Urban Plan. Dev.* **2017**, *143*, 04017007. [CrossRef]
33. Cavaignac, L.; Petiot, R. A quarter century of Data Envelopment Analysis applied to the transport sector: A bibliometric analysis. *Socio Econ. Plan. Sci.* **2017**, *57*, 84–96. [CrossRef]
34. He, Q.; Han, J.; Guan, D.; Mi, Z.; Zhao, H.; Zhang, Q. The comprehensive environmental efficiency of socioeconomic sectors in China: An analysis based on a non-separable bad output SBM. *J. Clean. Prod.* **2018**, *176*, 1091–1110. [CrossRef]
35. Mandl, U.; Dierx, A.; Ilzkovitz, F. *The Effectiveness and Efficiency of Public Spending*; (No. 301); Directorate General Economic and Financial Affairs; European Commission: Brussels, Belgium, 2008.
36. Litman, T.; Burwell, D. Issues in sustainable transportation. *Int. J. Glob. Environ. Issues* **2006**, *6*, 331–347. [CrossRef]
37. Lin, J.J.; Shin, T.Y. Does transit-oriented development affect metro ridership? Evidence from Taipei, Taiwan. *Transp. Res. Rec.* **2008**, *2063*, 149–158. [CrossRef]
38. Cook, W.D.; Liang, L.; Zhu, J. Measuring performance of two-stage network structures by DEA: A review and future perspective. *Omega* **2010**, *38*, 423–430. [CrossRef]
39. Seoul Metropolitan Council. *The Enforcement Rules of the Urban Planning Ordinance of Seoul*; Seoul Rules No. 4224; Legal Administration Service: Seoul, Korea, 2018.
40. Sung, H.; Oh, J.T. Transit-oriented development in a high-density city: Identifying its association with transit ridership in Seoul, Korea. *Cities* **2011**, *28*, 70–82. [CrossRef]
41. Sohn, K.; Shim, H. Factors generating boardings at metro stations in the Seoul metropolitan area. *Cities* **2010**, *27*, 358–368. [CrossRef]
42. Korea Energy Economics Institute and Korea Energy Agency. *Energy Consumption Survey*; Ministry of Trade, Industry and Energy: Sejong, Korea, 2017.

Article

Evaluation of Energy-Environment Efficiency of European Transport Sectors: Non-Radial DEA and TOPSIS Approach

Boban Djordjević and Evelin Krmac *

Faculty of Maritime Studies and Transport, University of Ljubljana, 6320 Portorož, Slovenia
* Correspondence: evelin.krmac@fpp.uni-lj.si

Received: 20 June 2019; Accepted: 26 July 2019; Published: 28 July 2019

Abstract: Transport is recognized as a major energy consumer and environment pollutant. Recently scholars have paid considerable attention to the evaluation of transport *energy and environmental efficiency* (*EEE*). In this paper, the non-radial Data Envelopment Analysis (DEA) model was employed to evaluate *EEE* on a macro level—i.e., of European road, rail and air sectors. The evaluation was conducted under the joint production framework, which considers energy and non-energy inputs, and desirable and undesirable outputs for the last ten years period. To rank decision-making units and check the aptness of this non-radial DEA model in transport *EEE* evaluation, the Technique for Order of Preference by Similarity to Ideal Solution (TOPSIS) method has been proposed. An empirical study has been conducted for as many European countries as possible, depending on availability of data. Based on the non-radial DEA model, it could be said that the level of *EEE* is improving for the road sector, while many evaluated countries have low *EEE* for the rail transport sector. Additionally, results have indicated that the TOPSIS method is more suitable than the non-radial DEA model in transport *EEE* evaluation and for identification of best practices.

Keywords: evaluation; energy; environment; efficiency; transport; DEA; TOPSIS

---

1. Introduction

1.1. Background

During recent decades, there have been increased debates concerning the gradual increase of global warming and the resulting climate change. The primary sources of global warming are increased concentrations of greenhouse gas (GHG) emissions, primarily carbon dioxide (CO_2), which is a product of human activities. The transport sector is one of the inevitable and essential parts of human activities, backbone of the economy, representing advantages for society in terms of transportation of goods and people, market integration, and provision of growth and jobs. It has been estimated that transport sector within the European Union (EU) contributes for 7% of European gross value added and 7.06% of employment [1].

Yet despite benefits, transport activities include disadvantages related to responsibilities for enormous energy consumption and resulting GHG emissions. According to the European Environmental Agency [2], with 348.5 Mtoe (Million tonnes oil equivalent), the transport sector was the biggest energy consumer in 2013, followed by households (295.9 Mtoe), industry (276.6 Mtoe), services (152.5 Mtoe) and fishing, agriculture, forestry and non-specified (30.2 Mtoe). Among transport modes in 2012, road transport had the largest share in the amount of consumed energy (307.5 Mtoe), followed by air (international and domestic) transport (51.5 Mtoe), international marine bunkers (46.4 Mtoe), rail transport (7.2 Mtoe), and domestic navigation (5.7 Mtoe) [1].

To ameliorate these disadvantages, the European Commission periodically published White Papers and emphasizing where the targets of EU policies were highlighted. The strategy set by the European Commission [3] was based on targets such as:

- Low emissions through reduction of 60% of GHG emissions by 2050 with respect to their 1990 level;
- Improvement of energy efficiency by decreasing final oil consumption and dependency ratio. The reduction was estimated at 12 to 13% by 2030 and to about 70% by 2050;
- Limited growth of congestion due to better multimodal solutions and new technologies.

Presently, the need for meeting the demands of transportation services and enhancing mobility is increasing, as well as the need for improving the *EEE* [1]. Awareness and concern about the energy consumption and environmental problems are becoming increasingly important worldwide. Numerous techniques have been employed to address the issues related to energy and the environment. The technique, which has received great attention, is the Data Envelopment Analysis (DEA) method as a non-parametric approach to efficiency evaluation [4]. Recognizing the share that transport has in energy and environmental problems, and having in mind the potential of the DEA method in *energy-environment efficiency* evaluation, DEA has been included in the analysis of transport *EEE*. The DEA method has been used in *EEE* analysis for different sectoral levels, countries and regional levels, as well as timely levels [5]. However, *EEE* evaluation and comparison of transport sectors on a macro level for EU countries is missing. Since the countries of the EU could have different strategies and measures in energy consumption and environment protection, it is of the utmost importance to identify the best practice.

1.2. The Aim and the Scope of the Paper

The aim of this paper is twofold. The first is to evaluate and analyze the changes of *EEE* of European road, air, and rail transport sectors, where the methodology for evaluating *EEE* is based on a non-radial DEA model proposed by Wu et al. [6] for 2006–2008, 2010, 2012, and 2014–2016, using the available data for the European countries which represent DMUs. The second aim of the paper is the introduction of the TOPSIS method in the evaluation of *EEE*, where the TOPSIS method is used for the ranking of DMUs. The evaluation of transport *EEE* has been done under the joint production framework, using non-energy inputs (labor and transport assets) and energy input (energy consumption) to produce desirable outputs (volume of passenger and freight transport) and undesirable output (GHG emissions). Aside from other widely used non-radial DEA models such as Slack-based models, Russell measure models, and Directional distance function, in this paper, the non-radial DEA model has been chosen due to its ability to use different non-proportional adjustments, with decision maker specified weights assigned to each efficiency score, and because of its ability to proportionally decrease the amounts of energy inputs and undesirable outputs simultaneously as much as possible [5,6].

The main contributions of this study are: (i) a newly systematic literature review in the field of transport *EEE* evaluation, (ii) a new definition of transport *EEE*, (iii) the evaluation of *EEE* with an extended set of used inputs, (iv) the evaluation of *EEE* of road, air, and rail transport sectors of European countries and their changing tendencies in terms of the *EEE*, (v) use of non-radial DEA and introduction of the TOPSIS method through DMUs ranking in the evaluation of transport *EEE*, as well as the comparison of their results and the identification of the most suitable one for the evaluation of the transport *EEE*. Based on the evaluation with the non-radial DEA model all stakeholders can create a sense of tendencies in terms of *EEE* of EU transport sectors. Through the introduction of the TOPSIS method for the same purpose, the science community can consider it as a potential tool for monitoring changes regarding *EEE*.

The following section presents the review of previous papers which have used DEA or TOPSIS methods in terms of transport *EEE* evaluation. Section 3 describes the methodology and considers which DEA model is appropriate for our purpose as well as the adoption of the TOPSIS method. The data used, DMUs selection, energy input, non-energy inputs, desirable outputs and undesirable output for EU countries are described in the second part of this section. Section 4 offers an overview of inputs and outputs for transport sectors and compares the results produced by non-radial DEA and the TOPSIS method, as well a discussion related to the obtained results. Finally, the summary of this study and some future directions in transport *EEE* evaluation are presented in Section 5.

2. Literature Review

The aim of the literature review was to perform an overview of papers related to the evaluation of energy efficiency and environment efficiency or both in the field of transport using different DEA models. In addition, a literature review was conducted as a basis for the process of identification of inputs and outputs for the non-radial DEA model. Moreover, a literature review was made in order to confirm the novelty of the introduction of the TOPSIS method in the evaluation and ranking of DMUs in *EEE*. Consequently, the literature review was focused on identifying the papers related to the evaluation and analysis of transport *EEE* with the DEA and TOPSIS methods, as well as in their combination.

The search strategy consisted of a literature review of relevant studies published in peer-reviewed journals within scientific sources such as Ebsco, ScienceDirect, Scopus, Springer, and Taylor and Francis, without limitation on the time period of publishing. The search, performed on titles, abstracts, and keywords for English written full-text free-available scientific journal papers, was finished in April 2019. Conference papers, projects, periodicals, and working papers related to this topic were not included in our review because they went through a less rigorous peer-review process. The application of keywords such as "energy efficiency AND Data Envelopment Analysis", and "environment efficiency AND Data Envelopment Analysis", "energy efficiency AND Technique for Order of Preference by Similarity to Ideal Solution", and "environment efficiency AND Technique for Order of Preference by Similarity to Ideal Solution", as well as the combinations where acronyms of methods were used, resulted in finding a large number of papers from various fields. To reduce this number, the reading of abstracts was performed and only the papers that analyzed energy or environment efficiency, and those that studied the application of the DEA method and the TOPSIS technique for the evaluation of one of the efficiencies, related to transport were extracted. In the second step, the reading of full texts of these papers was performed and finally, 35 relevant papers were extracted after removing duplicates.

In terms of the literature, for the evaluation of energy efficiency or environment efficiency, as well as the *EEE* evaluation different methods were used; such as-frequently used DEA methods, the Stochastic Frontier Model (SFA), and the TOPSIS method [7]. Judging by the number of papers reviewed in [4,5], it could be said that numerous studies used DEA for evaluation of energy efficiency or environment efficiency, as well as for *EEE* evaluation.

Initially, numerous papers dealing with the evaluation of energy efficiency considered energy consumption as input within a production framework without considering undesirable outputs. Four perspectives treating undesirable outputs could be found in the literature, such as: undesirable variables treated as inputs, undesirable measures treated by distinguishing between weak and strong disposability, integration of undesirable outputs into DEA models through the classification of invariance property where classifications of efficiencies and inefficiencies are invariant to the data transformation, and those where operational and environmental performance can be divided into two aspects using a measure of efficiency referred to as the range-adjusted measure [8]. Consequently, Zhou and Ang [9] proposed several DEA models within a joint production framework for energy efficiency evaluation, including undesirable outputs that were not considered in earlier proposed DEA models for energy efficiency evaluation.

Additionally, a considerable amount of studies employed DEA in transport *EEE* evaluation. Some papers applied DEA in transport energy efficiency or environment efficiency evaluation, while certain studies conducted the evaluation of transport *EEE*.

In this section, reviewed papers are categorized in terms of the used DEA models and the TOPSIS method, studied field (energy or environment efficiency, or *EEE*), inputs and outputs used in the evaluation (see Table 1), as well as in terms of definitions of energy efficiency, environment efficiency or *EEE*. Papers in which inputs and outputs were not classified as desirable and undesirable were classified separately in one special group.

2.1. Review of Methods and Techniques for Transport Energy Efficiency, Environment Efficiency, and EEE Evaluation

A large number of studies have presented extensions to basic DEA models such as incorporation of undesirable outputs, using efficiency measures (radial, non-radial, slack-based, hyperbolic, directional distance function), investigating changes in efficiency over time [4,5]. A radial DEA model has been used by Ramanathan [10] to compare the energy efficiency of rail and road transport in India, while in terms of the radial DEA model, Ramanathan [11] has presented an extended DEA model to estimate the energy consumption of the same modes of transport, resulting in a pre-specified DEA efficiency. Additionally, non-radial DEA models have been presented and have been used by Zhou and Ang [9] for measuring the energy efficiency performance of 21 OECD countries.

Different DEA models have already been proposed for *energy and environment*, as well as *energy-environment efficiency* evaluation. Regarding transport *EEE* evaluation, some authors have used traditional DEA models as a support tool for evaluating eco-efficiency for the different types of bioethanol transportation [12] and to evaluate the relative energy efficiency of rail, road, aviation and water transport [13]. Some models with particular modifications have been used for transport *EEE* analysis, such as radial and non-radial DEA models [8] taken from Zhou and Ang [9], a virtual frontier benevolent DEA cross efficiency model [14], a three-stage virtual frontier DEA model [15], a slack-based measure (SBM) DEA model [16,17], a non-radial SBM-DEA model [18–20], a parallel DEA approach [6], and parallel SBM-DEA model [21]. Furthermore, several papers have presented *EEE* evaluation in combination with other methods, such as an improved non-radial SBM-DEA model with window analysis [21] and Tobit regression, a super-efficiency SBM model with a window DEA model [22], bootstrapped data DEA-VRS models, DEA and directional distance functions to compute Leunberger productivity [23], economic input output life cycle assessment (EIO-LCA) and DEA by Egilmez and Park [24].

2.2. Review of Transport Energy Efficiency Evaluation

One of the first papers in road and rail transport energy efficiency evaluation and analysis of changes over time in India using DEA was presented by Ramanathan [10]. The presented approach was further extended by Ramanathan [11] in order to project energy consumption and estimate environmental efficiency for the periods 2005–2006 and 2020–2021. The transportation energy efficiency was evaluated by Cui and Li [15] for provincial administrative regions of China. Additionally, Zhou et al. [8] examined maximum energy-saving potential of the transport sector in 30 administrative regions of China. Moreover, the energy efficiency of 11 airlines was studied by Cui and Li [25]. Energy consumption by road, rail, aviation, and water transport modes using a DEA model and future transport energy consumption using an extended DEA model in China for the period from 1971 to 2011 were estimated by Lin et al. [13]. The transportation energy efficiency of Yangtze River Delta's 15 cities in the period from 2009 to 2013 has been studied by Chen et al. [26]. Then, Feng and Wang [27] have analyzed energy efficiency and the savings potential in China's transportation sector. Using DEA-cooperative game approach, Omrani et al. [28] have evaluated energy efficiency of transportation sector of 20 provinces in Iran.

2.3. Review of Transport Environment Efficiency Evaluation

The environmental efficiency of the transportation sector for 30 Chinese provinces was analyzed by Chang et al. [18]. The evaluation of the environmental performance for the transport industry was also elaborated upon by Beltrán-Esteve and Picazo-Tadeo [23]. Their study focused on changes in the environmental performance from eco-innovation and catching up with the best environmental technologies. An empirical study was conducted for 38 countries, including European, for the periods 1995–96 and 2008–09. Similarly, in terms of Europe, energy efficiency trends of five energy industries, including transport for 23 EU countries over the period 2000–2009 were evaluated by Makridou et al. [29] using DEA combined with the Malmquist productivity index. However, Hu and Honma [30] employed SFA in the evaluation of energy efficiency of OECD countries for 10 industries, including transport. Song et al. [31] presented a measurement of the environmental efficiency of highway transportation systems in 30 regions of China. The assessment of the environmental efficiency was conducted by Park et al. [19] through estimation of carbon efficiency and potential carbon reduction for 50 U.S. states. Additionally, Chang [20] analyzed the environmental efficiency of ports in Korea and estimated potential CO_2 emission reduction by ports in the country. Furthermore, Leal Jr et al. [12] evaluated eco-efficiency for chosen bioethanol transportation modes (roadway, railway, waterway, and pipeline) in Brazil. Some papers evaluated transport sectors in terms of several different viewpoints. Overall and individual environmental efficiency and resource use of 30 Chinese regional railway transport and highway transport subsectors were evaluated by Liu et al. [21]. Using SBM-DEA Chang and Zhang [32] have evaluated carbon efficiency of transportation sectors in China and Korea. In addition, with SBM-DEA model, Chu et al. [33] have analyzed environmental efficiency of transport systems. Chang et al. [17] studied environmental and economic efficiency of 27 global airlines. Analyzing impacts of the European Union Emission Trading Scheme (EU ETS) on airline performance was presented in [34]. Dynamic Environmental DEA was used for analyzing the impacts of 18 large global airlines from 2008 to 2014. Li et al. [35] conducted an analysis of impacts of included aviation into EU ETS on airline efficiency for 22 international airlines from 2008 to 2012 through three stages—i.e., operations, services and sales—using a Network Slacks-Based Measure with weak disposability and Network Slacks-Based Measure with strong disposability. Technical and environmental performance evaluation for major airlines from China, north Asia, and Europe over the period 2007–2010 was studied by Arjomandi and Seufert [36]. Egilmez and Park [24] quantified transportation related carbon, energy and water footprints of U.S. manufacturing sectors and evaluated environmental vs. economic performance based on eco-efficiency scores.

2.4. Review of Transport Energy-Environment Efficiency Evaluation

Regarding *energy-environment efficiency*, Wu et al. [6] measured energy and environment performance of passenger and freight transportation subsystems of 30 provincial-level regions in mainland China. The *energy-environmental efficiency* of road and railway sectors of 30 provinces in China was presented by Liu et al. [21]. Total factor *energy and environmental efficiency* of 30 of China's regional transportation sectors in terms of energy saving and CO_2 emission reduction were elaborated by Liu and Wu (2015).

Different non-energy and energy inputs, as well as desirable and undesirable outputs, were used in the process of *energy or environmental* and *energy-environment efficiency* evaluation with presented DEA models (Table 1).

Table 1. Inputs and outputs in transport *EEE* evaluation.

Author(s)	Sectors	Energy Inputs	Non-Energy Inputs	Desirable Outputs	Undesirable Outputs
Wu et al. [6]	passenger subsystem	energy consumption volume	passenger seats; capital; highway mileage	passenger turnover volume	CO_2 emissions
	freight subsystem	energy consumption volume	cargo tonnage, capital; highway mileage	freight turnover volume	CO_2 emissions
Zhou et al. [8]	/	million ton coal equivalence	labor	passenger kilometers; tons-kilometers	CO_2 emissions
Ramanathan [2,3]	rail, road	energy consumption	/	passenger kilometers; ton- kilometers	/

Table 1. *Cont.*

Author(s)	Sectors	Energy Inputs	Non-Energy Inputs	Desirable Outputs	Undesirable Outputs
Leal Jr. et al. [4]	road, rail, water, and pipeline	total energy consumption; atmospheric pollution; GHG emission; quantity of used lubricating oil discarded during maintenance	/	freight revenue received, the total cost of accidents	/
Lin et al. [5]	road, rail, aviation, and water	energy consumption	/	passenger kilometers; freight ton-kilometers	/
Cui and Li [6]	/	energy consumption volume	labor; capital	freight turnover volume; passenger turnover volume	/
Liu and Wu [7]	/	the volume of energy consumed	labor; capital	a value-added amount in the transportation sector	CO_2 emissions
Chang et al. [8]	/	the volume of energy consumed	labor; capital	GDP by transportation sector	CO_2 emissions
Park et al. [9]	/	energy consumption	capital expense; labor	value added (GDP)	CO_2 emissions
Chang [10]	ports	energy consumed	labor; capital	cargo tonnage; vessel tonnage	CO_2 emissions
Liu et al. [11]	railway	/	railway length; locomotives	passenge turnover; freight turnover	CO_2 emissions
	highway	/	highway length and automobiles	passenger turnover; freight turnover	CO_2 emissions
Cui and Li [12]	airline	tons of aviation kerosene	labor; capital	revenue ton kilometers; revenue passenger kilometers; total business income	CO_2 emissions
Chen et al. [13]	/	energy consumption	labor; capital	passenger volume and freight volume	carbon dioxide
Omrani et al. [14]	/	consumption volume of gasoline, oil gas and nature gas	labor; capital	GDP; passenger kilometers (PKM) and tone kilometers (TKM)	emission of greenhouse gases
Song et al. [15]	highway	gasoline consumption; diesel consumption	highway mileage; employed population	passenger capacity; passenger turnover; freight volume; freight turnover	nitrogen oxide; particulate matter emissions; the equivalent sound level of road noise
Chu et al. [16]	/	energy	labor; capital	value-added	CO_2 emissions
Arjomandi and Seufert [17]	airline	/	labor; capital	ton kilometres available (TKA)	CO_2 emissions (only for environmental efficiency model)
Cui et al. [18]	airline	aviation kerosene	number of employees	total revenue	greenhouse gas emission (GHG)

2.5. Review of Unclassified Inputs and Outputs

Moreover, some unreasonably classified and unsorted variables, such as *available seat kilometers* (ASK) with fuel consumption added as inputs, *revenue per ton kilometers* (RTK) as output and *carbon emissions* as undesirable output to estimate the environmental efficiency of airlines were employed in Chang et al. [17]. Cui and Li [37] evaluated the transportation carbon efficiency through inputs such as *carbon dioxide emissions*, *number of employees* in the transportation sector, and *transportation service import volume* for each selected country, while *freight and passenger turnover* volume were used as outputs. The evaluation was conducted with a virtual frontier DEA, while for the investigation of factors of the impact of carbon efficiency was made with Tobit regression. Cui et al. [38] evaluated factors that influence airline energy efficiency. The evaluation was performed using the Virtual Frontier Dynamic Slacks Based Measure, where the *number of employees* and *aviation kerosene* are used as the inputs, while *revenue ton kilometers*, *revenue passenger kilometers* and *total business income* are the outputs. The evaluation of impacts of including aviation into EU EST on airline efficiency, for each stage Li et al. [35] defined inputs and outputs. Within the operations stage, the *number of employees* and *aviation kerosene* were used as inputs, while *available seat kilometers* and *available ton kilometers* were used as outputs. For service stage, inputs were the *available seat kilometers, available ton kilometers* and *fleet size*, while outputs were the *revenue passenger kilometers* and *revenue ton kilometers* and undesirable output is *greenhouse gas emission* (the unique undesirable output). The *revenue passenger kilometers*, and the *revenue ton kilometers* and *sales costs* were inputs within the sales stage, while the *total business income* was output for this stage. In addition, through stages—i.e., operations and carbon abatement stages, Cui and Li [34] have evaluated the airline energy efficiency using Network SBM with weak disposability. *Salaries, wages and benefits, fuel expenses* and *total assets* were used as inputs within the operation stage, while *revenue passenger kilometers*, revenue *ton kilometers* and *estimated carbon dioxide* represented outputs. In carbon abatement, stage inputs were *estimated carbon dioxide* and *abatement expense*, while *carbon dioxide represented* the output. In measuring the energy efficiency of airlines Li et al. [35,39] Virtual Frontier Dynamic range adjusted measure was used, where the *number of employees* and *tons of aviation kerosene* represented inputs, while outputs were *revenue ton kilometers, revenue passenger kilometers*, and *total business income*.

In Beltrán-Esteve and Picazo-Tadeo [23] three environmental pressures—i.e., *global warming potential, tropospheric ozone formation potential* and *acidification potential* were used as inputs, while the *economic outcome* of the transport industry was used as an output which was measured using real gross output in purchasing parity power in evaluation environmental performance. The three environmental impact categories, i.e., *carbon footprint, water footprint* and *energy footprint* represented inputs, while a single output was *$/ton-km carriage*, used by Egilmez and Park [24] for evaluation of environmental vs. economic performance of manufacturing sectors.

2.6. Review of Application of TOPSIS Method for Transport EEE Evaluation

In the field of transport *EEE* evaluation, the real picture regarding the TOPSIS method is rather different compared to DEA. One could find a few studies where the TOPSIS method was employed in the field of the estimation of environmental efficiency of thermo power plants [40], decision making among various alternatives in eco-efficient chemical processes design [41], benchmarking building energy performance [42], selection of optimal solutions for energy consumption and thermal comfort [43], finding optimal solutions for district heating systems through various aspects such as fuel, temperature regime, level of building energy efficiency [44]. Moreover, Wang et al. [7] have used the TOPSIS method to analyze the overall hydropower efficiency in Canada from different points of view, which imply environment, technology, economy, benefits and social responsibility. However, the application of the TOPSIS method in the evaluation of transport *EEE* is not present in the literature.

2.7. Review of Definitions of EEE

Several papers have presented definitions of energy efficiency or environment/environmental efficiency. For example, eco-efficiency in Egilmez and Park [24] was defined as "the ratio of total economic activity in million dollars to the overall environmental impact". Transport energy in Cui and Li [15] was defined as "an efficiency, which is calculated by comparing the relationship between the outputs and the inputs". Additionally, Cui and Li [25] have considered energy efficiency for airlines as "the relationship between the outputs and the inputs". Environmental performance has been defined by Beltrán-Esteve and Picazo-Tadeo [23] as "the quotient between economic performance and ecological performance". Since the definitions of *EEE* of transport are missing in the reviewed papers, in this paper *energy-environment efficiency* of transport sectors is defined as the ratio of the total amount of energy consumption to production of GHG emissions as a result of the transportation process.

3. Methodology

DEA, as a type of multi-criteria decision analysis (MCDA) method, has mainly been applied for evaluation of relative efficiency. Additionally, it has been used as a benchmarking tool rather than choosing alternatives as the best solutions or directions in traditional decision making [4]. For measuring energy and environment efficiency, in the literature, radial and non-radial models are the two most widely used in DEA [21]. Radial DEA models proportionally decrease the amount of inputs and outputs, which may have weak discriminating power [6], lead to partial ranking in which most of the DMUs have the same score of efficiency [45], as well as occurrence of difficulties in ranking the environmental performance of efficient DMUs [20]. When including the environmental variable in the model, efficiency measuring is a challenging task because the environmental pollutant need not increase or decrease proportionally with outputs or inputs [19], and, consequently, non-radial DEA has a higher discriminating power than radial in the environmental performance thanks to non-proportional adjustments of different inputs/outputs in comparing DMUs [4]. Radial models also need to especially treat a negative or zero value in a data set; they do not have the property of "translation invariance" so cannot directly handle zero [46]. In addition, they do not provide information regarding the efficiency of the specific inputs and outputs included in the process [18,20,47]. To overcome such weaknesses, non-radial models have been developed are widely used in empirical research [21,48]. According to Lui et al. [21], the non-radial DEA model also causes less bias.

In this paper, a two-step methodology for the evaluation of transport *EEE* of EU countries has been employed. In the first step, the non-radial DEA model proposed by Wu et al. [6] has been used for the evaluation of transport *EEE*. The proposed non-radial model provides the use of decision makers' specified weights, different non-proportional adjustments, and proportionally decreases several energy inputs and undesirable outputs simultaneously to the degree possible. However, based on the fact that the DEA method, in the case of the same efficiency of two DMUs, cannot rank DMUs and provide evaluation of DMUs with simultaneously minimization and maximizations of inputs and outputs, we have proposed a Technique for Order of Preference by Similarity to Ideal Solution (TOPSIS) as an MCDA method for benchmarking the alternatives—i.e., decision making units (DMUs), detecting the best practices based on alternative rank and evaluation of transport *EEE*.

Hence, the TOPSIS method has been proposed to rank DMUs, and simultaneously compare efficiency scores vs. DEA results. Based on the content of TOPSIS—i.e., consideration of DMUs from different viewpoints (for example, through inputs and outputs that are presented as cost criterion and a beneficial criterion) this method was introduced for evaluation and ranking of DMUs for monitoring changes of *EEE*. Consequently, for this purpose the following research hypothesis was defined: Any similarity between the results of the evaluation and analyzing of *EEE* through the application of non-radial DEA model and TOPSIS method does not exist.

In these terms, questions that this paper endeavors to answer involve changes to *EEE* for EU transport sectors and the suitability and applicability of the TOPSIS method in the evaluation of *EEE*. Therefore, the objective of this paper is not to study factors of *EEE*, but rather to evaluate the *EEE* for EU transport sectors using the non-radial DEA model and consider the utility of the TOPSIS method regarding evaluation of *EEE*.

3.1. A Brief Description of DEA Method

The DEA method was proposed by Charnes et al. [49] and presents a non-parametric frontier approach for evaluating the relative efficiency of a set of entities, DMUs, with multiple inputs and outputs [9,10,50]. A major stated advantage of DEA is that it does not require prior assumptions regarding underlying functional relationships between inputs and outputs [4] and weights for input and output is calculated based on the input oriented Charnes, Cooper and Rhodes (CCR) DEA model [4] that can be written as: $\min\theta; s.t\ X\lambda \leq \theta x_i,\ Y\lambda \geq y_i,\ \lambda \geq 0$, where X and Y represent a set of vectors of inputs and outputs, respectively. θ represents a goal function of technical efficiency where $\theta \in [0,1]$. Based on the result, θ indicates how much an evaluated entity could potentially reduce its input vector while holding the output constant. The presented CCR model exhibits the constant returns to scale (CSR), but with additional constraint $\sum \lambda = 1$, the CCR model becomes the classical Banker, Chames and Cooper (BCC) model that allows the variant to return to scale (VRS) [4,51].

3.1.1. DEA Method for EEE Evaluation

DEA is strongly related to production theory, where raw materials and resources are treated as inputs, while products are treated as outputs in the production process [5,9]. Then, in the production process, in terms of evaluation of *energy and environmental efficiency*, desirable and undesirable outputs, are jointly produced by consuming both energy and non-energy inputs, where x, e, y and u are vectors of non-energy inputs, energy inputs, desirable outputs, and undesirable outputs, respectively. The joint production process can be represented as $T = \{(x, e, y, u); (x, e) \text{ can produce } (y, u)\}$.

Based on that let's assume that there are K DMUs, and each DMU uses n non-energy inputs and l energy inputs in order to produce m desirable outputs and j undesirable outputs denoted respectively as $x = (x_{1K}, \ldots, x_{nK})$, $e = (e_{1l}, \ldots, x_{LK})$, $y = (y_{mK}, \ldots, y_{mK})$, $u = \left(u_{1K}, \ldots, u_{JK}\right)$. Then, environment DEA production technology T exhibiting constant returns to scale (CRS) and incorporating undesirable outputs can be written as:

$$T = \{(x, e, y, u) : \sum_{k=1}^{K} \lambda_k x_{nk} \le x_n, n = 1, \ldots, N \tag{1}$$

$$\sum_{k=1}^{K} \lambda_k e_{lk} \le e_l, l = 1, \ldots, L, \tag{2}$$

$$\sum_{k=1}^{K} \lambda_k y_{mk} \ge y_m, m = 1, \ldots, M, \tag{3}$$

$$\sum_{k=1}^{K} \lambda_k u_{jk} = u_j, j = 1, \ldots, J, \tag{4}$$

where $\lambda_k \ge 0, k = 1, \ldots, K$.

Based on this, T reference technology, radial model, modified-radial, and non-radial models such as the Russell measure model, tone's slack-based model, range adjusted model and directional distance function model are used in energy efficiency and carbon emission efficiency in the literature. Additionally, there are four types of returns to scale (*RTS*) such as constant *RTS* (*CRS*) which is the most commonly used *RTS* category, non-increasing *RTS* (*NIRS*), non-decreasing *RTS* (*NDRS*) and variant *RTS* (*VRS*), where each of them reflects reference technology [5].

There are several DEA-type models, radial and non-radial, for pure energy efficiency evaluation with consideration of undesirable outputs, some of which can be used for estimating potential energy saving [9]. The radial model aims at reducing energy inputs as much as possible for the given level of non-energy inputs, plus desirable and undesirable outputs. Since the radial model has weak discriminating power in energy efficiency comparisons and does not consider energy mix effects, non-radial models for energy efficiency evaluation is also proposed in [8,9]. Therefore, the application of non-radial DEA models for energy efficiency evaluation considering undesirable outputs and maximized energy-saving potential, all under *CRS*, *NIRS* and *VRS* were presented in [8]. For example, if in the model (M) instead of limitation (5) we write $\sum_{k=1}^{K} \lambda_k \le 1$, $\sum_{k=1}^{K} \lambda_k \ge 1$ or $\sum_{k=1}^{K} \lambda_k = 1$, we receive non-radial model under *NIRS*, *NDRS*, and *VRS*, respectively. However, their non-radial models also attempt to reduce energy inputs as much as possible for the given level of non-energy input, desirable and undesirable outputs. In other words, their non-radial models do not consider reduction of undesirable outputs.

3.1.2. Non-Radial DEA Model for EEE Evaluation

Radial and non-radial DEA models for evaluating DMUs' total-factor *energy and environment efficiency* have been presented in Wu et al. [6]. To overcome all disadvantages of the presented radial model, following [52,53], in [6] the radial DEA model has been extended to the following non-radial model (M) for *energy-environment efficiency* evaluation as:

$$EEEI = min\ \frac{1}{2}(\frac{1}{L}\sum_{l=1}^{L}\theta_l + \frac{1}{J}\sum_{j=1}^{J}\theta_j) \tag{5}$$

s.t.

$$\sum_{k=1}^{K}\lambda_k x_{nk} \leq x_{n0}, n = 1, \ldots\ldots N \tag{6}$$

$$\sum_{k=1}^{K}\lambda_k e_{lk} \leq \theta_l e_{l0}, l = 1, \ldots\ldots L \tag{7}$$

$$\sum_{k=1}^{K}\lambda_k y_{mk} \geq y_{m0}, m = 1, \ldots M \tag{8}$$

$$\sum_{k=1}^{K}\lambda_k u_{jk} = \theta_j u_{j0},\ j = 1, \ldots\ldots J \tag{9}$$

$\lambda_k \geq 0, k = 1, \ldots, K.$

The model (M) will be used in this paper for *EEE evaluation* of EU transport sectors. The main advantage of the non-radial model (M) is that it proportionally decreases several energy inputs and undesirable outputs as much as possible for the given level of non-energy inputs and desirable outputs. The optimal values of *energy-environment efficiency index* (*EEEI*) are in the interval between 0 and 1. An entity with a higher value of *EEEI* has better *EEE* in terms of other entities. However, if the entity has *EEEI* equal to 1 it means that entity is the best, located on the frontier, and could not reduce energy input and undesirable output. Another benefit of the model is that (M) can consider energy input mix effects and undesirable outputs mix effects in the evaluation of *EEE* [6]. Such non-radial model (M) is suitable for *EEE* evaluation because it has a relatively strong discriminating power and capability to expand desirable outputs, simultaneously reducing undesirable outputs. Additionally, benefit lies in the fact that unified efficiency can be calculated through DM specified weights assigned to each of these two efficiency scores and depends on the preferences between energy use and environment protection performance. However, we have retained the weights as in the paper Wu et al. [6] and both are set to 1/2. These weights point to the similarity of the model (M) with TOPSIS method. Based on the all above pointed out simultaneous benefits in comparison to other non-radial DEA models and the fact that *EEE* evaluation in this paper couldn't be considered to be a dynamic change over time, we have chosen non-radial DEA model (M) by Wu et al. [6] for evaluating *energy-environment efficiency*.

3.2. Background of the TOPSIS Method

In this paper, the TOPSIS method proposed by Hwang and Yoon [54] has been employed as a decision-making tool to aid DMs in trade-off the whole DMUs. In the literature, this method has received much interest from researchers and practitioners that confirmed a wide range of real-world applications across different fields and specific sub-areas [55]. This method is based on the assumption that the selected alternative is to be at the least possible distance from the ideal positive solution and ideal negative solution. As one of the best and most frequently used methods, MCDM implies overall assessment, comparison, and ranking of alternatives. DEA divides DMUs into efficient and inefficient [49]. However, the question is, which of these efficient DMUs can be located in the higher position [56]. Based on that, it can be concluded that total discrimination of the DEA method can be low in some cases, especially in terms of differentiating efficient DMUs.

Therefore, our paper has included the TOPSIS method for finding the best alternative—i.e., for ranking and solving the drawbacks of the DEA method. Moreover, besides the fact regarding the great variety of existing DEA ranking methods, ranking DMUs such as cross-efficiency, super-efficiency, benchmarking, statistical techniques and so on, all consider DMUs only from one point of view—i.e., input-oriented or output-oriented views [56].

Consequently, an additional reason for selection of TOPSIS for *EEE* evaluation and ranking of DMUs is based on the content of TOPSIS—i.e., DM intention to rank DMU with the best ranking score closer to the positive ideal and to have the greatest distance from the negative ideal solution, and the ability of consideration of DMUs from both pessimistic and optimistic viewpoints—i.e., inputs and outputs, such as a cost and benefit criterion [56,57].

After application of DEA, the TOPSIS method was used to evaluate and rank DMUs to present the behavior of DMUs. For our purpose, the TOPSIS method has been employed for road, rail and air transport sectors following the steps in [7,43]:

1. Forming the decision matrix $X = \left[x_{ij}\right]_{n\times m}$; $i = 1,2,\ldots,n$; $j = 1,2,\ldots m$. Within the decision matrix, alternatives represent DMUs—i.e., European countries (Table 2), while for the criteria inputs and outputs for non-radial DEA model (Table 3) were chosen.
2. Normalization of decision matrix X in order to obtain normalized decision matrix $R = \left[r_{ij}\right]_{n\times m}$ by the vector normalization method that is presented as $r_{ij} = x_{ij}/\sqrt{\sum_{i=1}^{n} x_{ij}^2}$.
3. Calculation of the weight normalized decision matrix as $V = \left[v_{ij}\right]_{n\times m} = \left[w_j r_{ij}\right]_{n\times m}$, where w_j is a weight given to criteria from DM and sum of weights $\sum_{j=1}^{m} w_j = 1$. This method is appropriate for decision making which is based on criteria of different importance.

Table 2. EU countries and abbreviations.

DMUs-Countries
Belgium (BE), Bulgaria (BG), Czech Republic (CZ), Denmark (DK), Germany (DE), Estonia (EE), Ireland (IE), Greece (EL), Spain (ES), France (FR), Italy (IT), Cyprus (CY), Latvia (LV), Lithuania (LT), Luxembourg (LU), Hungary (HU), Malta (MT), Netherlands (NL), Austria (AT), Poland (PL), Portugal (PT), Romania (RO), Slovenia (SI), Slovakia (SK), Finland (FI), Sweden (SE), United Kingdom (UK), Croatia (HR)

Table 3. Variables for road, rail and air transport sectors.

Inputs/Outputs	Road	Rail	Unit	Air	Unit	Category
Labor	√	√	person in thousands	√	person in thousands	NEI_1 [1]
Number of assets	√	√	number in thousands	√	total	NEI [2]
Volume of energy consumption	√	√	Mtoe	√	Mtoe	EI [1]
Volume of freight transport	√	√	thousands mio pkm	√	thousands ton	DO_1 [3]
Volume of passenger transport	√	√	thousands mio pkm		million passengers	DO_2
GHG emissions	√	√	$MtCO_2e$ [4]	√	$MtCO_2e$	UDO [5]

[1] Non-energy input; [2] Energy input; [3] Desirable output; [4] Million ton of CO_2 equivalent; [5] Undesirable output.

In our paper, different weights have been delegated to each criterion for each transport sector. We have assigned the same weights to criteria for each year for the **road transport sector**, i.e., the *number of employees* ($w_i = 0.14$), *passenger cars* ($w_i = 0.15$), *freight vehicles* ($w_i = 0.15$), *energy consumed* ($w_i = 0.18$), *volume of passengers* ($w_i = 0.1$), *freight transport* ($w_i = 0.1$), and *GHG emissions* ($w_i = 0.18$).

The weights for criteria in the **rail transport sector** were the *number of employees* ($w_i = 0.16$), *total number of locomotives and railcars* ($w_i = 0.18$), y ($w_i = 0.2$), y ($w_i = 0.13$), *realized ton kilometers* ($w_i = 0.13$), and *GHG emissions* ($w_i = 0.2$). Finally, in the **air transport sector** we assigned the next weights to criteria: *number of employees* ($w_i = 0.18$), the *total number of aircraft by age* ($w_i = 0.16$), *energy consumed* ($w_i = 0.2$), *amount of transported goods* ($w_i = 0.13$), *number of transported passengers* ($w_i = 0.13$), and *GHG emissions* ($w_i = 0.2$).

4. Determination of positive ideal and negative ideal solutions is denoted as A^+ and A^-, respectively. In our case, A^+ and A^- represent the most efficient DMU and the most inefficient DMU, respectively, demonstrated as: $A^+ = \left\{\left(\begin{smallmatrix} max \\ i \end{smallmatrix} V_{ij}\middle| j \in J_+\right), \left(\begin{smallmatrix} min \\ i \end{smallmatrix} V_{ij}\middle| j \in J_-\right)\middle| i = 1,2,\ldots n\right\} = \left\{V_1^+,\ldots,V_m^+\right\}$ and $A^- = \left\{\left(\begin{smallmatrix} min \\ i \end{smallmatrix} V_{ij}\middle| j \in J_+\right), \left(\begin{smallmatrix} max \\ i \end{smallmatrix} V_{ij}\middle| j \in J_-\right)\middle| i = 1,2,\ldots n\right\} = \left\{V_1^-,\ldots,V_m^-\right\}$, where $J_+ =$

$(j = 1, 2, \ldots m)$ and $J_- = (j = 1, 2, \ldots m)$ are associated with benefit and cost criteria, respectively. In our research benefit criteria represent desirable outputs, while cost criteria include energy input, non-energy inputs and undesirable output (Table 3).

5. Calculation of the separation measure between each alternative by Euclidean distance. The separation of each alternative from the positive ideal is given as $S_i^+ = \sqrt{\sum_{j=1}^{m}\left(V_{ij} - V_j^+\right)^2}$, $i = 1, 2, \ldots n$, while the separation from the negative ideal is given as $S_i^- = \sqrt{\sum_{j=1}^{m}\left(V_{ij} - V_j^-\right)^2}$, $i = 1, 2, \ldots n$.
6. Calculation of the relative closeness A_i to the positive ideal solution A^+ defined as $C_i = S_i^+/\left(S_i^+ + S_i^-\right)$, $0 < C_i < 1$, $i = 1, 2, \ldots n$. If $C_i = 1$, it is clear that DMU is the most efficient, and if $C_i = 0$ then DMU is the most inefficient. DMU is closer to the most efficient as C_i approaches 1.
7. Ranking the alternatives—i.e., DMUs according to C_i, where a higher value of C_i denotes a better solution in terms transport *EEE*.

3.3. Selection of Data Set and DMUs

Energy-environment efficiency (*EEE*) of European road, rail and air transport sectors was examined. *EEE* of these transport sectors was analyzed for countries presented in Table 2.

Each country was defined as a DMU for conducting the non-radial DEA model. There were different rules of thumb for DMUs' number. According to Golany and Roll (1989) in order to make sure that the model was more discriminatory, the number of DMUs should be at least twice the number of inputs and outputs considered. Each of the DMUs was analyzed according to the road, rail and air transport sectors. DMUs were examined based on inputs and outputs represented in Table 3.

An empirical study was performed based on the available data set collected and compiled from "EU energy and transport in figures-statistical pocketbook" for 2006–2008, 2010, 2012–2018 [58–67]. However, only data for *a number of assets*, the *volume of passengers and freight transport* for air sector were combined with data from "Eurostat". This combination was made because the data for *the number of assets*, *volume of passengers and freight transport* for the air sector did not exist in the same form as the data for the road and rail sectors. For the air sector in the EU statistical pocketbooks, there is only the *volume of traffic* such as *revenue ton kilometers* and *revenue passenger kilometers* between member states, and similar data only for major airlines-but they are not represented for each country separately. The period of analyzing allowed us to track the changing trends in terms of *EEE* after the White Papers had been published. In case of absence of some data for energy input or undesirable output for particular DMU, the DMU was immediately eliminated from analysis. Consequently, in order to get reliable results, all numbers in the DEA had to be strictly positive (no zero values). This was mostly the case with the rail and air sectors.

During the application of DEA method, variables for outputs were chosen based on the research objective, while inputs were primarily resources used to generate outputs. However, it was essential to avoid exogenous variables which were not under the complete and direct control of DMUs [68].

Since the selection of inputs and outputs was a difficult task, we mainly chose them according to the literature review shown in Table 1. However, we added several new inputs, which were important in transport *EEE* analysis. Please note that presented inputs and outputs were used as a set of criteria in the application of the TOPSIS method. The inputs and outputs were selected for the road, rail, and air transport sectors in conducting non-radial DEA model and the TOPSIS method (Table 3). Their changes through selected time periods for each transport sector are described in Section 4 and can be seen in Figure 1a,b, Figure 2a,b and Figure 3. Based on the figures, comparison of transport sectors for each variable could be derived and it could be also determined, which one consumes minimum inputs and causes undesirable output for the realization of maximum desirable outputs.

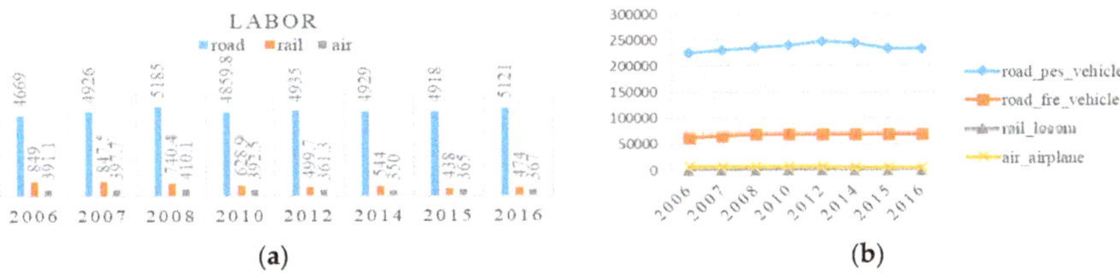

Figure 1. Trends of non-energy inputs for labor (**a**) and (**b**) number of assets.

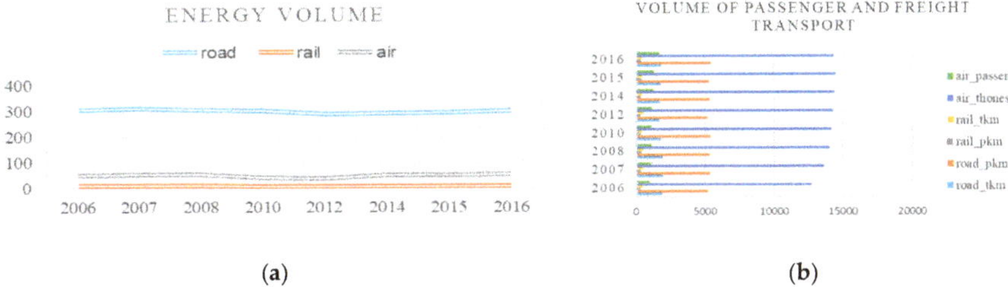

Figure 2. Trends of energy input (**a**), desirable outputs (**b**).

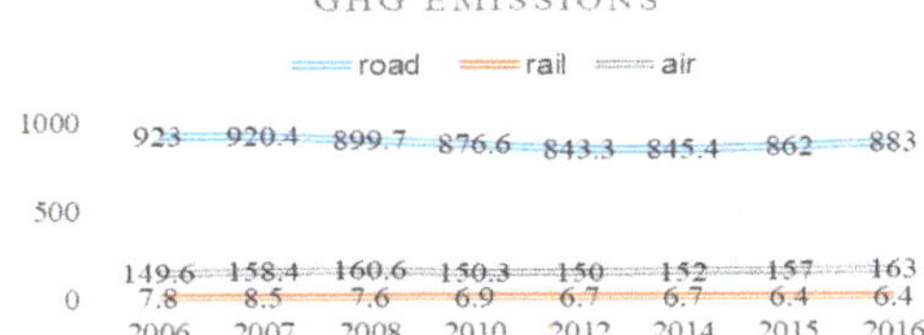

Figure 3. Undesirable output.

Non-energy inputs (*NEI*) for all sectors were the *number of assets* (see Table 4), and the *number of employees (labor)*. The *number of assets* represented the basic input to form the transport, the main energy consumers, and had a direct correlation with energy consumption. Therefore, we introduced them as non-energy inputs. Figure 1 represents the changes of labor (Figure 1a) and the number of assets (Figure 1b) represented as a sum for selected countries per each sector.

Table 4. Number of assets for road, rail and air transport.

Road transport	Passenger Vehicles: Stock of Registered Vehicles Including Buses, Coaches, and Passenger Cars
	Freight vehicles: good vehicles and powered two-wheelers
Rail transport	Total number of locomotives and railcars
Air transport	Total number of aircraft by age

Energy input (*EI*) represents the *amount of energy consumed* by each country per road, rail and air transport sectors expressed in million ton oil equivalent. Figure 2a shows the trend of energy consumption by each sector in terms of the selected period.

Desirable outputs (*DO*) involved a *volume of passengers and freight transport* (Figure 2b). For road transport sector volume of passenger transport represented a sum of realized kilometers by passenger cars, buses, and coaches, while the volume of freight transport consisted of realized national and international haulage. Regarding the rail transport sector, realized passenger and ton kilometers represented a volume of passenger and freight transport. In terms of air transport sector, the volume of passenger and freight transport represented the amount of transported goods and number of passengers, respectively.

Undesirable output (*UDO*) was the *total amount of greenhouse gas emissions* by chosen sector. Figure 3 shows the trends of undesirable output as a sum for all selected countries for all sectors.

4. Results and Discussion

4.1. Analysis of Inputs and Outputs

Figure 1, where labor and number of assets for the road, rail and air transport sectors in European countries were shown, indicated that the road sector had a dominant number of employees, followed by the air and rail transport sectors. Within the road sector, the number of employees increased year by year with an insignificant decrease in 2010 and the largest number in 2008. In rail transport the number of employees continuously decreased until 2012, after that it started do slightly increase, while in air transport there was an increase from 2006 to 2008, an decrease from 2010 to 2014, when the number of employees was the lowest, after that it started again slightly to increase. These trends of road and air transport could be the result of the economic crisis. However, almost constant reduction of the number of employees in the rail and air transport sectors could be the consequence of the intensive improvement of economic efficiency, which does not include any measurement regarding employment (or unemployment). Regarding the number of assets, the leading sector was again the road sector, primarily in terms of passenger vehicles. The road sector showed steady growth of several assets, while the rail sector highlights continuously decrease. Several assets gradually increased after 2006 in the air sector and it were slightly reduced in 2012.

Trends of energy input, desirable outputs and undesirable output are shown in Figure 2a,b and Figure 3, respectively. The volume of energy consumption in the rail sector was the lowest as compared to the road and air sectors. Energy volume was reduced in road, rail and air after 2008 and started again to increase after 2012. Reasons for decreasing energy consumption could be found in the increase of oil prices and the strategy of de-carbonization. These reasons were especially notable in air transport which had more dominant freight transport as compared to passenger transport, even though both showed constant growth. Please note that the volume of both types of transport in air sector was expressed in thousand ton and millions of passengers. The volume of rail passenger and freight transport was increasing up to 2008, after which it continued to dominate and showed growth in 2012. However, passenger transport was reduced. The volume of freight and passenger transport in road transport was reduced in 2008, after which the volume of freight transport decreased, while the volume of passenger transport was constantly dominant, with a slight reduction in 2012. As for GHG emissions, gradual reduction in road, rail, and air until 2014 could be noted, probably due to technological advances in vehicles and sources of energy, as well as more stringent standards [23]. After 2014 in road and air transport the volume of GHG emissions started to increase, while in rail it remained unchanged.

4.2. Results of DEA Method

In this part of the paper, based on the objective of the study, the results of the application of the non-radial DEA model are presented. At this point, the potential factors of *EEE* are only mentioned, without any statistical or other analysis.

All DEA results were calculated by Excel Solver. The calculation was conducted for each transport sector separately and for each year. The availability of data was the best in terms of road transport sector, followed by air and rail.

The results for the **road transport sector** (Tables 5 and 6) indicated that the best *EEEI* for countries was in the green cells each year, meaning that these countries were relatively energy-environment efficient. The countries in red cells with the least *EEEI* were: Cyprus (CY) in 2006, 2008, 2010, 2012, 2014–2016, and Austria (AT) in 2007. Please note that *EEEI* was improved for both countries in 2012 and 2014 as compared to previous years, but after 2014 it slightly deteriorated. Regarding Cyprus (CY) it can be seen (see supplementary material) that in these years, all values of data for desirable variables were lower, while those for undesirable were higher in comparison with other DMUs. However, Austria (AT) was inefficient, probably due to a higher number of undesirable variables in comparison to other DMUs (see raw data in supplementary material). The improvement of *EEE* in the road sector could be a result of stricter policy measures through prioritization in de-carbonization with primary introduction of CO_2 emission standards for new passenger cars and heavy vehicles [1,50], highlighted the use of bioenergy and renewable energy [62], new technologies for vehicles and traffic management [1], as well as improved conditions of cabotage. The best value of *EEEI* for each year of the evaluation was for Lithuania (LT) and Luxemburg (LU), almost each year for Slovakia (SK) and Slovenia (SI) and thus they could be considered the countries with the best practices. The main reason lies in the fact that these countries had the lowest values for undesirable variables in comparison with other DMUs, while desirable variables were higher and comparable with other DMUs (see supplementary material). It could be also seen that most of the countries improved their *EEEI* in the period of 2014–2016, while Ireland (IE) worsened drastically the values of *EEEI* after 2012.

Table 5. Results of efficiency of non-radial DEA model and rank of TOPSIS method for the road sector (2006–2012).

	Road Sector									
	2006		2007		2008		2010		2012	
DMUs	DEA Non-Radial	TOPSIS Rank	DEA Non-Radial	TOPSIS Rank	DEA Non-Radial	TOPSIS Rank	DEA Non-Radial	TOPSIS Rank	DEA Non-Radial	TOPSIS Rank
BE	0.808649	17	0.929032	17	0.319853	19	0.769692	16	1	19
BG	0.546216	11	0.844309	8	0.357289	10	0.958431	9	1	7
CZ	0.637093	15	0.591504	16	0.631525	15	0.658219	18	0.612003	15
DK	1	13	1	14	0.343223	14	0.705374	13	0.91319	12
DE	1	26	1	25	0.444898	25	0.81077	25	1	28
EE	0.617846	5	0.622808	1	0.686869	6	0.708056	5	0.807294	5
IE	1	12	1	13	0.312296	13	1	12	1	14
EL	0.625318	20	0.717725	20	0.288885	20	0.81967	20	1	21
ES	0.564873	24	0.570622	24	0.607714	24	0.538116	23	0.742671	24
FR	0.86936	25	0.821412	26	0.325536	26	1	26	1	26
IT	1	27	1	27	0.289674	27	0.918007	27	0.997767	27
CY	0.419031	10	0.421868	10	0.17257	8	0.353261	8	0.493888	10
LV	0.780133	3	0.776361	5	0.741368	4	1	4	1	4
LT	1	1	1	3	1	2	1	2	1	1
LU	1	9	1	9	1	9	1	7	1	11
HU	0.550362	14	0.612014	12	0.623609	12	0.810149	10	0.941513	9
MT	0.747098	7	0.685189	7	0.091075	5	0.651429	6	0.669092	6
NL	0.605102	21	0.614736	21	0.511777	21	0.535244	21	0.681301	22
AT	0.43343	19	0.409689	19	0.398355	18	0.440404	19	0.490857	20
PL	0.769045	22	0.869976	22	0.697239	22	0.954807	22	0.993453	23
PT	0.560112	18	0.70439	15	0.446918	17	0.61483	17	0.792474	16
RO	1	6	1	4	0.808434	11	0.765552	14	0.78381	17
SI	1	4	1	6	0.753484	3	1	3	1	3
SK	1	2	1	2	1	1	1	1	1	2
FI	1	8	1	11	0.525652	7	0.887975	11	0.917381	13
SE	0.536462	16	0.605111	18	0.415428	16	0.875713	15	0.967949	18
UK	1	23	1	23	0.268397	23	0.878957	24	1	25
HR	/	/	/	/	/	/	/	/	0.744544	8

Green color: the best *EEEI*. Red color: the least *EEEI*.

Table 6. Results of efficiency of non-radial DEA model and rank of TOPSIS method for the road sector (2014–2016).

Road Sector						
	2014		2015		2016	
DMUs	DEA Non-Radial	TOPSIS Rank	DEA Non-Radial	TOPSIS Rank	DEA Non-Radial	TOPSIS Rank
BE	1	22	1	24	1	23
BG	1	9	1	10	1	9
CZ	0.735	17	0.762	16	0.705	14
DK	0.705	11	0.705	13	0.709	11
DE	1	25	1	27	1	26
EE	0.841	15	0.918	17	0.867	16
IE	0.769	8	0.716	8	0.695	7
EL	1	7	1	4	1	3
ES	0.732	24	1	25	1	24
FR	1	27	0.932	28	0.948	28
IT	1	26	1	26	1	25
CY	0.549	14	0.523	12	0.475	13
LV	0.898	16	0.944	18	0.895	15
LT	1	2	1	2	1	1
LU	1	10	1	9	1	8
HU	0.924	12	0.879	14	0.854	12
MT	0.675	6	0.622	6	0.691	6
NL	1	23	1	22	1	20
AT	0.571	18	0.554	23	0.554	22
PL	1	20	1	15	1	17
PT	0.853	5	0.808	11	0.837	10
RO	0.813	21	0.901	20	0.906	19
SI	1	1	1	3	1	2
SK	1	4	1	7	1	4
FI	0.902	13	1	19	0.710	18
SE	0.995	19	0.929	21	0.990	21
UK	0.868	28	1	1	1	27
HR	0.768	3	0.721	5	0.698	5

Green color: the best *EEEI*. Red color: the least *EEEI*.

As far as the **rail transport sector** was concerned, the number of DMUs was smaller due to data unavailability (Tables 7 and 8). It could be noticed that the number of units with the highest value of *EEEI* was in 2006. The most efficient countries were represented in green cells per year. The least value of *EEE* Index in 2006, 2007, 2008, and period 2014–2016 was in a red cell for Greece (EL), due to lack of data, the second one for 2010 and 2012 were the United Kingdom (UK) and Romania (RO). However, similar to the case with road transport, inefficiency of these countries can be related to higher values of undesirable variables while desirable variables were lower in comparison with other DMUs (see supplementary material). It would be interesting to note that Latvia (LV), Italy (IT) and Sweden (SE) (data available only for 2006–2008, 2010, and 2012) had a constant best value of *EEEI* and represented the best practices. Based on the supplementary material, i.e., raw data, it can be seen that these countries had lower values for *energy and GHG emissions* while values for *volume of passenger and freight transport* were high in comparison with other DMUs. In terms of countries considered per each year, it could be concluded that scores of efficiency were not homogeneous. In 2012 half of DMUs were improved, while the other half of DMUs deteriorated. In the period 2014–2016 it could be noted drastically improvement of *EEEI* for Germany (DE), Austria (AT), and Poland (PL). In terms of the rail sector, the value of efficiency scores declined and a decline in efficiency for some countries could be attributed to insufficient market opening and modernization of rail sectors, incomplete implementation of modern traffic management systems such as ERTMS for European railway, insufficient European high speed rail network and interoperability, lack of modal shift in each country—i.e., involvement in the transport market [1,3]-as well as incomplete electrification of railway networks.

Table 7. Results of efficiency of non-radial DEA model and rank of TOPSIS method for the rail sector (2006–2012).

	Rail Sector									
	2006		**2007**		**2008**		**2010**		**2012**	
DMUs	**DEA Non-Radial**	**TOPSIS Rank**	**DEA Non-Radial**	**TOPSIS Rank**	**DEA Non-Radial**	**TOPSIS Rank**	**DEA Non-Radial**	**TOPSIS Rank**	**DEA Non-Radial**	**TOPSIS Rank**
BE	0.771182	10	0.593661	9	0.716647	9	1	8	1	5
BG	0.490449	11	0.45795	8	0.309252	10	/	/	/	/
CZ	0.474833	16	0.467384	14	0.336824	13	0.391971	15	0.271762	10
DK	0.971713	8	0.899017	10	0.637045	11	0.413052	11	0.625863	8
DE	0.836313	19	0.857799	20	0.651819	19	0.653537	1	0.913981	15
EE	1	4	/	/	/	/	0.371963	13	/	/
IE	/	/	/	/	/	/	/	/	/	/
EL	0.170862	9	0.177281	11	0.158075	12	/	/	/	/
ES	0.666978	15	0.618655	16	0.454679	16	0.347673	17	0.528261	12
FR	1	18	1	19	1	18	0.948728	3	1	14
IT	1	17	1	17	1	14	1	4	/	/
CY	/	/	/	/	/	/	/	/	/	/
LV	1	6	1	4	1	3	1	6	1	4
LT	0.901956	5	0.984458	5	0.833261	5	0.712766	9	0.115761	7
LU	/	/	/	/	/	/	/	/	/	/
HU	1	12	0.57653	13	0.465083	7	0.323137	14	1	1
MT	/	/	/	/	/	/	/	/	/	/
NL	1	13	1	6	1	6	0.651163	10	0.963769	9
AT	0.951996	2	0.689355	2	0.563669	2	0.661709	7	0.879885	3
PL	1	16	0.933333	18	0.85144	17	0.777128	12	0.395634	11
PT	0.519374	7	0.670578	7	0.433445	8	/	/	/	/
RO	0.754073	14	0.394416	15	0.288133	15	0.329787	16	0.10985	13
SI	/	/	/	/	/	/	/	/	/	/
SK	/	/	0.830108	12	/	/	/	/	/	/
FI	0.980779	3	0.897204	3	0.705479	4	1	5	0.348291	6
SE	1	1	1	1	1	1	1	2	1	2
UK	0.593252	20	0.518208	21	0.382882	20	0.289564	18	0.607946	16
HR	/	/	/	/	/	/	/	/	/	/

Green color: the best *EEEI*. Red color: the least *EEEI*.

Table 8. Results of efficiency of non-radial DEA model and rank of TOPSIS method for the rail sector (2014–2016).

	Rail Sector					
	2014		2015		2016	
DMUs	DEA	TOPSIS	DEA	TOPSIS	DEA	TOPSIS
	Non-Radial	Rank	Non-Radial	Rank	Non-Radial	Rank
BE	1	5	0.468	10	0.442	9
BG	/	/	0.221	16	/	/
CZ	0.335	11	0.418	9	0.439	8
DK	0.295	16	0.329	17	0.317	18
DE	1	19	1	20	1	20
EE	0.5	8	0.5	14	0.5	13
IE	0.5	17	0.5	18	0.5	16
EL	0.049	15	0.075	15	0.090	14
ES	1	6	0.620	4	1	6
FR	0.559	20	1	12	1	17
IT	1	2	1	6	1	4
CY	/	/	/	/	/	/
LV	1	9	1	7	1	7
LT	0.591	12	0.609	11	0.689	10
LU	/	/	/	/	/	/
HU	0.410	7	0.605	5	0.763	5
MT	/	/	/	/	/	/
NL	0.357	10	0.587	3	0.603	3
AT	1	1	1	1	1	1
PL	1	3	1	8	1	2
PT	/	/	/	/	/	/
RO	0.201	13	0.273	15	0.299	12
SI	/	/	/	/	/	/
SK	0.5	4	0.5	2	0.5	15
FI	0.623	18	0.633	19	0.762	19
SE	/	/	/	/	/	/
UK	0.291	21	0.351	21	0.358	21
HR	0.181	14	0.132	13	0.131	11

Green color: the best *EEEI*. Red color: the least *EEEI*.

For the **air transport sector**, the availability of data was better than in the rail sector (Tables 9 and 10), and the *EEEI* was also better compared to rail. The highest values of *EEEI* were for countries Cyprus (CY) and Luxembourg (LU). They had the best scores of efficiency throughout the entire evaluation period. Belgium (BE) and the Netherlands (NL) had the best value until 2012, after that their *EEE* indices drastically decreased. The lowest value of *EEEI* was in red cells for the United Kingdom (UK) in 2006, followed by Finland (FI) in 2007, the United Kingdom (UK) in 2008 and 2010, Portugal (PT) in 2012, Ireland (IE) in 2014 and 2015, and France (FR) in 2016. Similar to previous modes of transport, DMUs with higher values of desirable variables and lower values of undesirable variables (see supplementary material) in comparison with other DMUs have better values of *EEEI*.

Surprisingly, the United Kingdom (UK) with three red values until 2012, had the best values for all three last years of evaluation period. The inefficiency of DMUs could be attributed to old aircraft, waiting for improvement of their aircraft's fuel efficiency, or switching to green fuels [36].

Table 9. Results of efficiency of non-radial DEA model and rank of TOPSIS method for the air sector (2006–2012).

	Air Sector									
	2006		2007		2008		2010		2012	
DMUs	DEA Non-Radial	TOPSIS Rank	DEA Non-Radial	TOPSIS Rank	DEA Non-Radial	TOPSIS Rank	DEA Non-Radial	TOPSIS Rank	DEA Non-Radial	TOPSIS Rank
BE	1	1	1	1	1	2	1	2	1	1
BG	/	/	0.706722	12	0.65753	10	0.821158	7	0.82482	11
CZ	0.980766	10	0.83411	9	0.827037	19	1	6	0.890042	6
DK	0.859713	14	0.69759	15	0.937466	3	0.897875	10	0.612392	14
DE	0.768173	19	0.642743	22	0.696089	23	1	19	0.751725	23
EE	/	/	0.71923	4	/	/	/	/	/	/
IE	1	12	1	14	0.829558	16	1	14	0.844267	19
EL	/	/	1	11	1	14	1	11	1	3
ES	1	18	0.946155	20	1	22	1	20	0.865467	21
FR	0.672936	20	0.625645	21	0.64646	24	0.580559	21	0.641927	24
IT	1	17	0.87757	19	0.924698	21	0.842596	18	0.878126	20
CY	1	3	1	5	1	4	1	3	1	7
LV	1	7	0.764706	7	0.888889	9	0.977778	5	0.923833	9
LT	0.729208	5	0.661475	6	1	7	/	/	0.842346	10
LU	1	2	1	2	1	1	1	1	1	2
HU	1	4	1	3	0.812856	8	1	4	1	4
MT	0.938662	6	/	/	1	5	/	/	0.703901	13
NL	1	16	1	18	1	20	1	17	1	15
AT	0.945716	13	0.833572	17	0.829785	17	0.916049	13	0.795946	16
PL	1	9	1	10	1	11	1	8	0.902043	8
PT	0.742749	15	0.629629	16	0.609771	18	0.647523	15	0.567826	17
RO	1	8	0.988549	8	0.844111	12	0.670554	9	/	/
SI	/	/	/	/	/	/	/	/	/	/
SK	/	/	/	/	1	6	/	/	/	/
FI	0.733084	11	0.566175	13	0.60486	15	0.609107	12	0.70449	5
SE	/	/	/	/	0.939663	13	0.869726	16	1	18
UK	0.648914	21	0.571663	23	0.574808	25	0.500327	22	1	22
HR	/	/	/	/	/	/	/	/	0.766475	12

Green color: the best *EEEI*. Red color: the least *EEEI*.

Observing the highest values of *EEEI* for all transport sectors, Luxembourg (LU) was most frequently present in road and air transport sector, while data for the Luxembourg rail transport sector were missing. United Kingdom (UK) showed the lowest values of *EEEI* for rail and air transport sector.

4.3. Results of the TOPSIS Method

As with any other method, DEA also has its drawbacks. Regardless of its orientation, the DEA method has a tendency to assign maximum or minimum values to input and output, regardless of their initial values, by assigning the best value for *EEEI*. To eliminate this problem, weights of TOPSIS were used for considering the initial values of input and output variables. Furthermore, non-radial DEA shows discriminating power but does not indicate the difference between DMUs with efficiency results of 1. Consequently, a defect in the DEA analysis is the existence of multiple efficient units. In the literature, different DEA ranking methods exist for ranking DMUs that attempt to consider DMUs from input or output oriented aspects.

Table 10. Results of efficiency of non-radial DEA model and rank of TOPSIS method for the air sector (2014–2016).

Air Sector						
	2014		2015		2016	
DMUs	DEA Non-Radial	TOPSIS Rank	DEA Non-Radial	TOPSIS Rank	DEA Non-Radial	TOPSIS Rank
BE	0.459	22	0.404	22	0.064	22
BG	0.759	13	0.621	11	1	3
CZ	0.843	6	0.757	7	0.225	6
DK	0.697	2	0.554	9	1	1
DE	1	26	1	26	1	26
EE	0.5	11	0.5	12	0.5	10
IE	0.299	23	0.267	23	0.196	23
EL	1	9	1	10	0.144	17
ES	0.821	25	0.761	25	0.147	25
FR	0.417	27	0.439	27	0.063	28
IT	0.932	20	0.928	18	0.229	18
CY	1	7	1	6	1	8
LV	0.680	8	0.541	14	0.217	15
LT	0.656	14	0.594	13	0.210	16
LU	1	1	1	1	1	2
HU	1	3	1	4	0.384	5
MT	0.436	17	0.531	19	0.200	19
NL	0.355	24	0.382	24	0.055	24
AT	0.443	18	0.307	20	0.254	20
PL	0.952	5	0.969	5	0.214	7
PT	0.506	21	0.499	21	0.148	21
RO	1	12	0.734	15	0.197	12
SI	0.324	15	0.332	17	0.346	14
SK	0.5	10	0.5	8	0.243	9
FI	0.633	19	0.605	2	0.219	4
SE	0.836	4	0.822	3	0.179	11
UK	1	28	1	28	1	27
HR	0.858	16	1	16	0.217	13

Green color: the best *EEEI*. Red color: the least *EEEI*.

Therefore, the TOPSIS method with both viewpoints—i.e., pessimistic and optimistic-was used in order to evaluate and rank DMUs. Moreover, TOPSIS was employed with the aim of checking the results of the non-radial DEA model. Based on all these considerations, in order to verify differences between these two methods a research hypothesis was formed. The results of the TOPSIS method were calculated using Excel environment.

In terms of the **road sector** one country ranked first in three years, Lithuania (LT) in 2006, 2012 and 2016, while Slovakia (SK) ranked first in two years, 2008 and 2010. In 2007 the best ranked was Estonia (EE), in 2014 Slovenia (SI), and in 2015 United Kingdom (UK). In all cases, the *EEEI* was 1 (see Tables 5 and 6).

In the **rail sector**, Sweden (SE) received a rank of 1 in 2006, 2007 and 2008, and Austria (AT) in 2014, 2015, and 2016. Germany (DE) and Hungary (HU) were ranked first in 2010 and 2012 (see Tables 7 and 8). In all cases, except in the case of Germany (DE), the *IEEE* was 1.

As for **air transport**, Belgium (BE) was ranked 1 in 2006, 2007, and 2012, Luxembourg (LU) in 2008, 2010, 2014, and 2015, while Denmark (DK) received rank of 1 in 2016. In all cases the *EEEI* was 1. (see Tables 9 and 10).

All the countries with a rank of 1 for the rail and air transport sectors (except the DE in 2010 for the rail sector) at the same time had the best value of *EEEI*. However, the results of TOPSIS method were different. For instance, for the road sector Lithuania (LT) was ranked 1 by TOPSIS in 2006, 2012 and 2016 and also had the best *EEEI* for those years, as well as Slovakia (SK) in 2008 and 2010, while in 2007 Estonia (EE) whose *EEEI* was 0.622 had a rank of 1.

In addition, the results of the TOPSIS method were different from the results of the non-radial DEA model. Estonia (EE), for example, had a rank of 1 in 2007 even though the result of the *EEEI* of the DEA model was lower: 0.622. Furthermore, considering other DMUs, we note similar situations. For the road sector in 2012, DMUs with ranks from 1 to 4 obtained from TOPSIS had efficiency scores of 1 obtained by the non-radial DEA model, while DMU with rank 5 had an efficiency score of 0.807. Moreover, for the same year Belgium (BE) with a rank of 19 by TOPSIS had an efficiency score 1 by the non-radial DEA method. The situation is similar for other years; for example, Luxembourg (LU) had an efficiency score of 1 for 2008 and Ireland for 2010, while with TOPSIS Luxembourg (LU) had 9 and Ireland (IE) 12. From 2014 to 2016 Belgium (BE) has *EEEI* equal 1, but it was ranked as 22, 24, and 23 respectively. The similar is for Bulgaria (BG), Germany (DE), Italy (IT), Luxemburg (LU), Netherlands (NL), and Poland (PL).

It is significant to note that the results of the TOPSIS method for the rail and air transport sectors were different for a large number of DMUs in comparison to the non-radial DEA model. For example, for rail Sweden (SE) was ranked first in 2006, 2007 and 2008, while in 2012 the best ranked was Hungary (HU); on the other side, both DMUs had the highest efficiency scores. However, in 2010 Germany (DE) was ranked first, although by the non-radial DEA model the obtained efficiency score was 0.653537. Germany (DE), Italy (IT), Latvia (LV), and Poland (PL) had the efficiency scores for 2014, 2015, and 2016 equal 1, while they were not ranked as first by TOPSIS.

Similar to the results of the TOPSIS for road, for rail France (FR) received an efficiency score of 1 in 2006 and 2008, yet was ranked 18; and for 2007 and 2012 it ranked 19 and 14 while having the highest efficiency score.

Regarding the air sector, the picture in terms of results given by DEA and TOPSIS is the same as with the road and rail sectors. Belgium (BE), with an efficiency score of 1 in 2006, 2007 and 2012 had a rank of 1, while in 2008 and 2010 Luxembourg (LU), with the highest efficiency score, ranked first. However, for example, Spain (ES), with an efficiency score of 1 by DEA model in 2006, 2008 and 2010 had ranks of 18, 22, and 20, while in 2007 the Netherlands (NL) ranked 18 with a 1 efficiency score, and in 2012 the United Kingdom (UK) ranked 22 yet had the highest efficiency score. Furthermore, Germany (DE) with the highest efficiency scores in 2014, 2015, and 2016 ranked 26.

Therefore, it could be said that the DEA is not the most suitable benchmarking tool in the field of the evaluation of the transport *EEE*.

Consequently, based on the significant differences between the results of the non-radial DEA model and the TOPSIS method, our research hypothesis could be confirmed. The reason for differences in results should be found in the fact that DEA considered inputs for a given level of outputs, while the TOPSIS method, in order to find the best DMUs, closest to the ideal positive solution and furthest from the negative weights its criteria. Another reason for differences in results of the TOPSIS method and the DEA method is the involvement of weights for each criterion, not only for variables in the goal function in the non-radial DEA model.

4.4. Discussion

Within Tables 5 and 6, the results of non-radial DEA model and TOPSIS method for **road sector** were presented. Based on the results of non-radial DEA model efficient and non-efficient DMU can be seen. Considering the results of evaluation through the selected period, it can be seen that the lowest number of countries with the efficiency score of 1 was in 2008. Numerous DMUs are efficient, while one of them has the lowest score of efficiency. However, due to discrimination power of non-radial DEA model there is a little difference between non-efficient DMUs. In addition, it can be noticed that many countries obtained the efficiency score of 1 by DEA model, while only one was ranked as first with TOPSIS. In 2007, the country ranked as first by TOPSIS received relatively low efficiency score—i.e., only 7 countries were less efficient. However, considering the raw data (see supplementary material) the main reason for that is related to the TOPSIS method and values of raw data. For EE (Estonia) in 2007 the values of data are lower than that used as minimum criteria in TOPSIS and other data as maximum criteria for that country were sufficient in comparison with other alternatives. In general, the efficiency scores of more than three quarters of evaluated countries constantly increased after the 2012. Only one country was inefficient throughout the entire evaluation period by DEA, while it was ranked among first half of all countries by TOPSIS. The primarily reason for the difference between results of DEA and TOPSIS lies in the fact that TOPSIS evaluates countries with different criteria from two points of view. However, with the DEA method it is possible to change the efficiency of some DMUs if the raw data for them is changed. Based on that, some inefficient DMUs can become efficient and vice versa.

Considering the **rail sector** (Tables 7 and 8) in comparison with road sector it can be seen that a significantly smaller number of countries have the highest efficiency. Within the rail sector, the most efficient countries were in 2006 and 2016. Regarding the application of TOPSIS method similar picture appears as in road sector. Beside the best efficiency score obtained with non-radial DEA model, some countries were near to the worst ranked by TOPSIS. Only one country had the efficiency score constantly very low throughout the entire evaluation period, and at the same time it was mostly ranked at the bottom of the list by TOPSIS.

Regarding the **air sector** (Tables 9 and 10) it can be seen that the number of countries that obtained the highest efficiency score by DEA method was greater in comparison with rail and road sectors. However, such results were obtained due to the highest volume of transport realized by air sector with the lowest number of used assets. Furthermore, the level of consumed energy and produced emissions were lower in comparison with road sector.

5. Conclusions

Over the last decade, the main and intensive topics of research among scholars were energy consumption and environmental impacts caused by transport systems. One of the major contributors to energy consumption and endangerment of the environment in Europe has been the overall transport sector. Among all modes of transport, the road sector was recognized as the main energy consumer and environmental pollutant. Notwithstanding the importance of this fact, there was not any research on *energy-environment efficiency* of European transport sectors.

In this paper, therefore *energy-environment efficiency* (*EEE*) of European road, rail and air transport sectors were evaluated using a modified non-radial DEA model under the joint production framework proposed by Wu et al. [6]. The evaluation was conducted for European countries in terms of road, rail and air transport sectors for the period 2006 to 2008, 2010, 2012, 2014, 2015, and 2016. The first reason for the adoption of non-radial DEA model was simultaneous minimization of energy inputs and undesirable outputs for the given level of inputs and outputs, and this was a primary motivation for our paper. This non-radial DEA model has benefits in terms of the ability to use different non-proportional adjustments and weighting for energy inputs and undesirable outputs. In the paper, non-energy inputs, named several assets (see Table 3), were defined and used in the evaluation of transport *EEE* for the first time.

Furthermore, the concept of transport *EEE* was introduced in this study through the reflection of the relationship among transport energy, non-energy inputs, and transport desirable and undesirable outputs. Following the aims of the paper, all used variables were described and their changes were presented only through figures-without any statistical analysis, while factors of *EEE* were only mentioned.

An additional contribution provided in the paper was the introduction of the TOPSIS method as a tool in the evaluation of transport *EEE* through the ranking of DMUs. With this evaluation of *EEE* for European road, rail, and air transport sectors, the stakeholders from each member state may find the best practices toward the most efficient means of improving overall efficiency.

Based on the results of the DEA approach, we found that the lowest number of DMUs with the best value of *EEEI* for the road sector was in 2008. In terms of rail transport, the highest DMUs had the best *EEEI* in 2006, and after a decrease in 2007, has since remained fairly unchanged. As far as air transport was concerned, the best value of *EEEI* was attributed to the least number of DMUs in 2007 and 2012.

Rail and air transport had much more room for *EEE* improvement than the road transport sector, which was relatively efficient in many European countries. Accordingly, it could be concluded that periodical documents of EU policies for sustainable transport contributed to the improvement of *EEE* in road transport sector. However, a modal shift as one of the policies and advanced technologies was not fully completed for rail transport. Therefore, the potential of the rail transport sector was not totally realized, which resulted in inefficiency within the rail transport sector. Ramanathan's [11] findings confirmed this, stating that rail transport could capture around 50% of the expected traffic, which would result in saving of about 37% in energy consumption and associated CO_2 emissions that would result if the existing patterns of modal split did not change. Additionally, Song et al. [22] stated that a higher rate of railway concentration was associated with higher environment efficiency. In terms of air transport, the measures for *EEE* improvement implied newer and more fuel-efficient aircraft through new technology and larger planes [36].

The main conclusions could be drawn through the application of the TOPSIS method. All DMU with *EEEI* result 1 had the first rank. However, in some cases DMUs with an *EEEI* score 1 and lower had a rather wildly varying ranks. This is because the non-radial DEA model minimizes desirable and undesirable inputs for a given level of the desirable outputs. Then, the non-radial model benchmarks one DMU in comparison with others DMUs. However, based on the changes of raw data (see supplementary material) with the non-radial DEA model some inefficient DMUs can become efficient and vice versa. Then, the consequence could be a result of the TOPSIS method considering all inputs and outputs with the possibility of minimization and maximization during the process of analysis–they strove for clear values. Furthermore, the weights used in the TOPSIS method were assigned to each input and output.

The authors proposed using the TOPSIS method for finding the best practice in accordance with the challenges of European transport. The main European challenge is the demand for transport, which has significantly increased since 2000 and is expected to continue growing. On the other hand, the European transport sector is heavily dependent on oil. It releases GHGs and air pollutants into the atmosphere and contributes to climate change, but also makes the European economy more vulnerable regarding fluctuations in global energy supplies and prices [69]. The overall improvement of transport *EEE* in Europe could be achieved through progress in terms of *EEE* for each member state for each transport sector.

Bearing that in mind, finding the best practice which realizes the highest volume of freight and passenger transport with minimal energy consumption and environmental impact could be found through the TOPSIS method rather than any DEA approach.

Consequently, the authors highlight the importance of including of the TOPSIS method in future evaluation of transport *EEE*. Some proposals for the development of the European transport sector in terms of *energy-environment efficiency* are:

I. Intensifying efforts in the implementation policy of modal shift from road and air transport sectors to eco-friendly sectors, such as rail transport, primarily in developed countries, in order to increase total *EEE*.
II. Strengthening transport infrastructure and infrastructure components in terms of rail transport at bottlenecks, as well as total modernization of rail transport sectors.
III. Reinforcing the adoption of technological innovations and standards in each transport sector.
IV. Employment of alternative sources of energy and modes of transport that have a potential to reduce energy consumption and environment impacts.

As for future work, the focus should be on research in terms of changes in results of non-radial DEA models with weights assigned to all inputs and outputs, as compared to the TOPSIS method. Additionally, TOPSIS could be used with other DEA models for checking results during the evaluation of transport *EEE*. Moreover, attention should be drawn to research into the impacts of technological innovation for improving transport *EEE*, primarily in the rail transport sector.

Supplementary Materials: The following are available online at http://www.mdpi.com/1996-1073/12/15/2907/s1, Table S1: Real data.

Author Contributions: Conceptualization, B.D. and E.K.; methodology, B.D. and E.K.; validation and formal analysis, B.D.; investigation, B.D. and E.K.; resources, and data curation, B.D.; writing, review and editing, E.K.; visualization, supervision, and funding acquisition, E.K. Both authors have read and approved the final manuscript.

Funding: This research received no external funding.

Conflicts of Interest: The authors declare no conflict of interest.

References

1. European Commission. *A European Strategy for Low-Emission Mobility-COM (2016) 501 Final*; European Commission: Brussels, Belgium, 2016.
2. European Environmental Agency. *Final Energy Consumption by Sector and Fuel*; European Environmental Agency: Brussels, Belgium, 2015.
3. European Commission. *White Paper on transport—Roadmap to a Single European Transport Area—Towards a Competitive and Resource-Efficient Transport System*; European Commission: Brussels, Belgium, 2011.
4. Zhou, P.; Ang, B.W.; Poh, K.L. A survey of data envelopment analysis in energy and environmental studies. *Eur. J. Oper. Res.* **2008**, *189*, 1–18. [CrossRef]
5. Meng, F.Y.; Su, B.; Thomson, E.; Zhou, D.Q.; Zhou, P. Measuring China's regional energy and carbon emission efficiency with DEA models: A survey. *Appl. Energy* **2016**, *183*, 1–21. [CrossRef]
6. Wu, J.; Zhu, Q.; Yin, P.; Song, M. Measuring energy and environmental performance for regions in China by using DEA-based Malmquist indices. *Oper. Res. Int. J.* **2015**, *17*, 715–735. [CrossRef]
7. Wang, B.; Nistor, I.; Murty, T.; Wei, Y.M. Efficiency assessment of hydroelectric power plants in Canada: A multi criteria decision making approach. *Energy Econ.* **2014**, *46*, 112–121. [CrossRef]
8. Zhou, G.H.; Chung, W.; Zhang, Y.X. Measuring energy efficiency performance of China's transport sector: A data envelopment analysis approach. *Expert Syst. Appl.* **2014**, *41*, 709–722. [CrossRef]
9. Zhou, P.; Ang, B.W. Linear programming models for measuring economy-wide energy efficiency performance. *Energy Policy* **2008**, *36*, 2911–2916. [CrossRef]
10. Ramanathan, R. A holistic approach to compare energy efficiencies of different transport modes. *Energy Policy* **2000**, *28*, 743–747. [CrossRef]
11. Ramanathan, R. Estimating energy consumption of transport modes in India using DEA and application to energy and environmental policy. *J. Oper. Res. Soc.* **2005**, *56*, 732–737. [CrossRef]
12. Leal, I.C.; de Almada Garcia, P.A.; de Almeida D'Agosto, M. A data envelopment analysis approach to choose transport modes based on eco-efficiency. *Environ. Dev. Sustain.* **2012**, *14*, 767–781. [CrossRef]

13. Lin, W.B.; Chen, B.; Xie, L.N.; Pan, H.R. Estimating Energy Consumption of Transport Modes in China Using DEA. *Sustainability* **2015**, *7*, 4225–4239. [CrossRef]
14. Cui, Q.; Li, Y. An empirical study on the influencing factors of transportation carbon efficiency: Evidences from fifteen countries. *Appl. Energy* **2015**, *141*, 209–217. [CrossRef]
15. Cui, Q.; Li, Y. The evaluation of transportation energy efficiency: An application of three-stage virtual frontier DEA. *Transp. Res. Part D Transp. Environ.* **2014**, *29*, 1–11. [CrossRef]
16. Liu, X.H.; Wu, J. Energy and environmental efficiency analysis of China's regional transportation sectors: A slack-based DEA approach. *Energy Syst. Optim. Model. Simul. Econ. Asp.* **2017**, *8*, 747–759. [CrossRef]
17. Chang, Y.T.; Park, H.S.; Jeong, J.B.; Lee, J.W. Evaluating economic and environmental efficiency of global airlines: A SBM-DEA approach. *Transp. Res. Part D Transp. Environ.* **2014**, *27*, 46–50. [CrossRef]
18. Chang, Y.T.; Zhang, N.; Danao, D.; Zhang, N. Environmental efficiency analysis of transportation system in China: A non-radial DEA approach. *Energy Policy* **2013**, *58*, 277–283. [CrossRef]
19. Park, Y.S.; Lim, S.H.; Egilmez, G.; Szmerekovsky, J. Environmental efficiency assessment of US transport sector: A slack-based data envelopment analysis approach. *Transp. Res. Part D Transp. Environ.* **2018**, *61*, 152–164. [CrossRef]
20. Chang, Y.T. Environmental efficiency of ports: A Data Envelopment Analysis approach. *Marit. Policy Manag.* **2013**, *40*, 467–478. [CrossRef]
21. Liu, Z.; Qin, C.X.; Zhang, Y.J. The energy-environment efficiency of road and railway sectors in China: Evidence from the provincial level. *Ecol. Indic.* **2016**, *69*, 559–570. [CrossRef]
22. Song, M.L.; Zhang, G.J.; Zeng, W.X.; Liu, J.H.; Fang, K.N. Railway transportation and environmental efficiency in China. *Transp. Res. Part D Transp. Environ.* **2016**, *48*, 488–498. [CrossRef]
23. Beltran-Esteve, M.; Picazo-Tadeo, A.J. Assessing environmental performance trends in the transport industry: Eco-innovation or catching-up? *Energy Econ.* **2015**, *51*, 570–580. [CrossRef]
24. Egilmez, G.; Park, Y.S. Transportation related carbon, energy and water footprint analysis of US manufacturing: An eco-efficiency assessment. *Transp. Res. Part D Transp. Environ.* **2014**, *32*, 143–159. [CrossRef]
25. Cui, Q.; Li, Y. Evaluating energy efficiency for airlines: An application of VFB-DEA. *J. Air Transp. Manag.* **2015**, *44–45*, 34–41. [CrossRef]
26. Chen, X.H.; Gao, Y.Y.; An, Q.X.; Wang, Z.R.; Neralic, L. Energy efficiency measurement of Chinese Yangtze River Delta's cities transportation: A DEAwindow analysis approach. *Energy Effic.* **2018**, *11*, 1941–1953. [CrossRef]
27. Feng, C.; Wang, M. Analysis of energy efficiency in China's transportation sector. *Renew. Sustain. Energy Rev.* **2018**, *94*, 565–575. [CrossRef]
28. Omrani, H.; Shafaat, K.; Alizadeh, A. Integrated data envelopment analysis and cooperative game for evaluating energy efficiency of transportation sector: A case of Iran. *Ann. Oper. Res.* **2019**, *274*, 471–499. [CrossRef]
29. Makridou, G.; Andriosopoulos, K.; Doumpos, M.; Zopounidis, C. Measuring the efficiency of energy-intensive industries across European countries. *Energy Policy* **2016**, *88*, 573–583. [CrossRef]
30. Hu, J.L.; Honma, S. A Comparative Study of Energy Efficiency of OECD Countries: An Application of the Stochastic Frontier Analysis. *Energy Procedia* **2014**, *61*, 2280–2283. [CrossRef]
31. Song, M.L.; Zheng, W.P.; Wang, Z.Y. Environmental efficiency and energy consumption of highway transportation systems in China. *Int. J. Prod. Econ.* **2016**, *181*, 441–449. [CrossRef]
32. Chang, Y.T.; Zhang, N. Environmental efficiency of transportation sectors in China and Korea. *Marit. Econ. Logist.* **2017**, *19*, 68–93. [CrossRef]
33. Chu, J.-F.; Wu, J.; Song, M.-L. An SBM-DEA model with parallel computing design for environmental efficiency evaluation in the big data context: A transportation system application. *Ann. Oper. Res.* **2018**, *270*, 105–124. [CrossRef]
34. Cui, Q.; Li, Y. Airline energy efficiency measures considering carbon abatement: A new strategic framework. *Transp. Res. Part D Transp. Environ.* **2016**, *49*, 246–258. [CrossRef]
35. Li, Y.; Wang, Y.Z.; Cui, Q. Energy efficiency measures for airlines: An application of virtual frontier dynamic range adjusted measure. *J. Renew. Sustain. Energy* **2016**, *8*, 015901. [CrossRef]
36. Arjomandi, A.; Seufert, J.H. An evaluation of the world's major airlines' technical and environmental performance. *Econ. Model.* **2014**, *41*, 133–144. [CrossRef]

37. Cui, Q.; Wei, Y.M.; Li, Y. Exploring the impacts of the EU ETS emission limits on airline performance via the Dynamic Environmental DEA approach. *Appl. Energy* **2016**, *183*, 984–994. [CrossRef]
38. Cui, Q.; Li, Y.; Yu, C.L.; Wei, T.M. Evaluating energy efficiency for airlines: An application of Virtual Frontier Dynamic Slacks Based Measure. *Energy* **2016**, *113*, 1231–1240. [CrossRef]
39. Li, Y.; Wang, Y.Z.; Cui, Q. Has airline efficiency affected by the inclusion of aviation into European Union Emission Trading Scheme? Evidences from 22 airlines during 2008–2012. *Energy* **2016**, *96*, 8–22. [CrossRef]
40. Montanari, R. Environmental efficiency analysis for enel thermo-power plants. *J. Clean. Prod.* **2004**, *12*, 403–414. [CrossRef]
41. Ouattara, A.; Pibouleau, L.; Azzaro-Pantel, C.; Domenech, S.; Baudet, P.; Yao, B. Economic and environmental strategies for process design. *Comput. Chem. Eng.* **2012**, *36*, 174–188. [CrossRef]
42. Wang, Y.; Chen, Y.; Benitez-Amado, J. How information technology influences environmental performance: Empirical evidence from China. *Int. J. Inf. Manag.* **2015**, *35*, 160–170. [CrossRef]
43. Delgarm, N.; Sajadi, B.; Delgarm, S. Multi-objective optimization of building energy performance and indoor thermal comfort: A new method using artificial bee colony (ABC). *Energy Build.* **2016**, *131*, 42–53. [CrossRef]
44. Ziemele, J.; Pakere, I.; Blumberga, D. The future competitiveness of the non-Emissions Trading Scheme district heating systems in the Baltic States. *Appl. Energy* **2016**, *162*, 1579–1585. [CrossRef]
45. Babazadeh, R.; Razmi, J.; Rabbani, M.; Pishvaee, M.S. An integrated data envelopment analysise mathematical programming approach to strategic biodiesel supply chain network design problem. *J. Clean. Prod.* **2017**, *147*, 694–707. [CrossRef]
46. Sueyoshi, T.; Goto, M. Returns to Scale and Damages to Scale with Strong Complementary Slackness Conditions in DEA Assessment: Japanese Corporate Effort on Environment Protection. *Energy Econ.* **2012**, *34*, 1422–1434. [CrossRef]
47. Choi, Y.; Zhang, N.; Zhou, P. Efficiency and abatement costs of energy-related CO2 emissions in China: A slacks-based efficiency measure. *Appl. Energy* **2012**, *98*, 198–208. [CrossRef]
48. Yaqubi, M.; Shahraki, J.; Sabouni, M.S. On dealing with the pollution costs in agriculture: A case study of paddy fields. *Sci. Total Environ.* **2016**, *556*, 310–318. [CrossRef]
49. Charnes, A.; Cooper, W.W.; Rhodes, E. Measuring the efficiency of decision making units. *Eur. J. Oper. Res.* **1978**, *2*, 15. [CrossRef]
50. Voltes-Dorta, A.; Perdiguero, J.; Jimenez, J.L. Are car manufacturers on the way to reduce CO2 emissions? A DEA approach. *Energy Econ.* **2013**, *38*, 77–86. [CrossRef]
51. Banker, R.D.; Charnes, A.; Cooper, W.W. Some Models for Estimating Technical and Scale Inefficiencies in Data Envelopment Analysis. *Manag. Sci.* **1984**, *30*, 1078–1092. [CrossRef]
52. Bian, Y.W.; Yang, F. Resource and environment efficiency analysis of provinces in China: A DEA approach based on Shannon's entropy. *Energy Policy* **2010**, *38*, 1909–1917. [CrossRef]
53. Wang, J.-M.; Sun, Y.-F. The Application of Multi-Level Fuzzy Comprehensive Evaluation Method in Technical and Economic Evaluation of Distribution Network. In Proceedings of the 2010 International Conference on Management and Service Science (MASS), Wuhan, China, 24–26 August 2010; pp. 1–4.
54. Hwang, C.-L.; Yoon, K. *Multiple Attribute Decision Making: Methods and Applications—A State-of-the-Art Survey*; Springer: Berlin, Germany, 1981.
55. Behzadian, M.; Otaghsara, S.K.; Yazdani, M.; Ignatius, J. A state-of the-art survey of TOPSIS applications. *Expert Syst. Appl.* **2012**, *39*, 13051–13069. [CrossRef]
56. Hosseinzadeh Lotfi, F.; Fallahnejad, R.; Navidi, N. Ranking Efficient Units in DEA by Using TOPSIS Method. *Appl. Math. Sci.* **2011**, *5*, 10.
57. Jahantigh, M.; Hosseinzadeh Lotfi, F.; Moghaddas, Z. TRanking of DMUs by using TOPSIS and different ranking models in DEA. *Int. J. Ind. Math.* **2013**, *5*, 9.
58. European Union. *EU Energy and Transport in Figures—Statistical Pocketbook 2009*; Publications Office of the European Union: Luxembourg, 2009.
59. European Union. *EU Energy and Transport in Figures—Statistical Packetbook 2010*; Publications Office of the European Union: Luxembourg, 2010.
60. European Union. *EU Transport in Figures—Statistical Pocketbook 2011*; Publications Office of the European Union: Luxembourg, 2011.
61. European Union. *EU Transport in Figures—Statistical Pocketbook 2012*; Publications Office of the European Union: Luxembourg, 2012.

62. European Environmental Agency. *Towards a Green Economy in Europe—EU Environmental Policy Targets and Objectives 2010–2050*; Publications Office of the European Union: Luxembourg, 2013.
63. European Union. *EU Transport in Figures—Statistical Pocketbook 2014*; Publications Office of the European Union: Luxembourg, 2014.
64. European Union. *EU Transport in Figures—Statistical Pocketbook 2015*; Publications Office of the European Union: Luxembourg, 2015.
65. European Union. *EU Transport in Figures—Statistical Pocketbook 2016*; Publications Office of the European Union: Luxembourg, 2016.
66. European Union. *EU Transport in Figures—Statistical Pocketbook 2017*; Publications Office of the European Union: Luxembourg, 2017.
67. European Union. *EU Transport in Figures—Statistical Pocketbook 2018*; Publications Office of the European Union: Luxembourg, 2018.
68. Nakanishi, Y.J.; Falcocchio, J.C. Performance assessment of intelligent transportation systems using data envelopment analysis. *Res. Transp. Econ.* **2004**, *8*, 181–197. [CrossRef]
69. European Environmental Agency. *Towards Clean and Smart Mobility—Transport and Environment in Europe*; Publications Office of the European Union: Luxembourg, 2016.

Article

A Two-Phase Method to Assess the Sustainability of Water Companies

Fátima Pérez *, Laura Delgado-Antequera and Trinidad Gómez

Departamento de Economía Aplicada, Universidad de Málaga, 29016 Málaga, Spain
* Correspondence: f_perez@uma.es

Received: 29 May 2019; Accepted: 5 July 2019; Published: 9 July 2019

Abstract: Composite indicators are becoming more relevant for evaluating the performance of water companies from a holistic perspective. Some of them are related with economic aspects, and others focus on social and environmental features. Consequently, a multidimensional evaluation is necessary for handling the great amount of information provided by multiple single indicators of a different nature. This paper presents a two-phase approach to evaluate the sustainability of water companies. First, a partial composite indicator for each dimension (social, environmental, economic) is obtained using multi-criteria decision making (MCDM). Then, a global indicator is obtained, in terms of the values reached in the previous stage for every partial indicator, by means an optimization problem rooted in data envelopment analysis (DEA). Our proposal offers the possibility of analyzing the performance of each water company under each dimension that characterizes the concept of sustainability, as well as a joint assessment including all the dimensions, facilitating the decision-making process. We apply it to evaluate the sustainability of 163 Portuguese water companies. The results show the strengths and weaknesses of each unit and serve as a guideline to decision-makers on the aspects for improving the performance of water utilities.

Keywords: composite indicator; sustainability; water utilities management; data envelopment analysis; multi-criteria decision making (MCDM)

1. Introduction

The evolution of water management is a key issue for the human development. An effective performance of such service is a challenge for the community. Designing a good management system requires considering different factors. In countries such as England and Wales, Portugal, Chile, or the Netherlands, the water industry exists as a monopoly, so that companies and administrations invest their efforts on comparing the different processes within the industry. In general, benchmarking is widely considered a good strategy to control and supervise the performance of this service. Ref. [1] provide a rigorous evaluation of the growing number of benchmarking studies dealing with performance scores based on production or cost estimates. At the same time, the literature reveals frequent use of performance indicators (PIs) when dealing with benchmarking, because of the multiple benefits it brings to the administrations, for instance, to contrast the regulatory conditions, compare, and/or evaluate the quality of the service and establish fair tariff policies. So, in order to control these values, water utilities-following industry regulations- provide systematic reports on different PIs to the government or administrators. The information delivered within this data includes management, environmental, financial and, more recently, social aspects related to water operations. However, different reasons make this set of indicators difficult to interpret because they do not offer a holistic view, as they do not reflect a measure of general performance.

To overcome this difficulty, a common approach is to aggregate the PIs into a unique indicator, named a composite indicator (CI). Although the literature offers a wide range of techniques to create a

CI, most of them use methodologies from multi-criteria decision analysis (MCDA). They have been used to develop CIs applied to diverse sectors of services, activities, or processes [2–4]. In particular, methodologies based on goal programming (GP) are of great interest for the construction of CIs and they have been successfully applied to diverse fields as tourism [5,6], manufacturing [7], human sustainable development [8–10], or environmental sustainability [11,12]. The main advantages of using GP to develop CIs are: it is not necessary to normalize the initial set of PIs; the CI uses the complete information included in the initial set of PIs; and it does not require a large number of units in comparison with the number of initial indicators.

Usually, another technique used to create CIs is data envelopment analysis (DEA) [13]. DEA is a linear programming tool for evaluating the performance of a set of peer entities that use one or more inputs to produce one or more outputs. As pointed out by [14], the main advantages of using DEA to construct CIs are: it provides a measure of performance based on real data; DEA models do not require the normalization of the initial data; and DEA respects the individual characteristics of the units and their own particular value systems. Techniques based on DEA have been developed to create CIs in [6,15–17].

Since the 1990s, governments of many countries and organizations have emphasized the importance of the concept of sustainability [18]. There is no consensus on the definition of this concept, although it is widely agreed that it must incorporate social, environmental, and economic factors which are interconnected ([19,20]). The water industry has not ignored this trend and, currently, it has extensively recognized its important role in establishing and operating sustainable water supplies and wastewater treatment systems [2,21]. There is clearly a need for a paradigm shift in the water companies, considering social and environmental aspects in the decision making process, not just economic issues [22,23]. In the framework of evaluating the sustainability of water companies, most of the literature focuses on evaluating the sustainability of physical and engineering aspects [24–26], from an environmental perspective [27] or economic sustainability [28,29]. However, there is a lack interest on assessing the sustainability of water companies themselves. In particular, only a few papers apply different techniques from MCDA to assess the sustainability of water companies from a multidimensional perspective. For instance, Ref. [30] construct an index by aggregating the PIs as a linear combination of their normalized values. Also, the MACBETH (Measuring Attractiveness by a Categorical Based Evaluation Technique [31]) method is used to evaluate the sustainability of water supply systems [24]. Another example, Ref. [19], applies the ELECTRE TRI-Nc (Elimination and Choice Expressing Reality [32]) method as a tool to integrate the dimensions of a quality of service index. Additionally, Ref. [12] combines the PIs using an index based on distance-principal components and another based on GP.

In view of the above, in this work, a method to assess the sustainability of the water companies is conducted, using the traditional approach of sustainability, which considers three dimensions into this concept: social, environmental, and economic. Then, a two-phase method combining GP and DEA is proposed, in order to take advantage of both methodologies. A similar two-phase method is proposed in [6] to evaluate the sustainability of Cuban nature-based tourism destinations. Nevertheless, in that work, the distance-principal component (DPC) composite indicator developed by [33] is used to sum up the initial PIs into the dimensions established (social, economic and patrimonial (Although it is usual to use "environmental dimension", in [6] it is replaced by "patrimonial dimension".) instead of GP. Choosing a technique based on GP comes from their good properties, as previously mentioned.

Then, in the first phase, a technique based on GP [5] is used to obtain the dimensional or partial CIs. In the field of water treatments, there is a lack of consensus on the appropriate criteria to select, in order to determine which PIs are involved in evaluating the status of water sustainability. Then, to overcome this difficulty, as suggested by [34], our proposal groups the initial indicators into the dimensions that characterize the concept of sustainability: social, economic, and environmental dimensions. In this way, when the first phase is applied, three-dimensional composite indicators (social, environmental, and economic) are obtained for each water company. This allows for independently analyzing the performance of each water company among these three dimensions.

Later, in the second phase, the dimensional indicators have to be aggregated in order to design a global composite indicator for evaluating the water companies' sustainability. At this point, a controversial question is the assignment of weights to each dimensional indicator. On the one hand, under some circumstances, it is not easy to obtain information from specialists to determine these weights. On the other hand, the assignation of the same weighting values for all the water companies could be complicated, as each of them might have their own particularities in terms of preferences. To overcome these issues, we have chosen, in the second phase, a DEA-based model known as "Benefit-of-the-Doubts" [32]. To do this, the values obtained in the previous stage are used as outputs of this "Benefit-of-the-Doubt" approach.

This two-phase approach offers the possibility of considering the strengths and weaknesses of each water company, as well as providing the decision-makers with useful information.

The hypothesis behind this study is that the water companies should manage their activity in a way as balanced as possible, from social, environmental, and economic point of view. In this sense, the approach proposed in this work allows evaluating and comparing the performance of water companies for each sustainability dimension and, later, identifying if such dimensions have or not a similar influence on the global score. This aspect is an advantage of the proposed approach in comparison to other procedures. In the first phase, an indicator is obtained for each sustainability dimension, and in the second phase the different sustainability dimension indicators are aggregated to build a global indicator. In this aggregation, the weights of the different dimensional indicators are endogenously determined using a DEA-based model, allowing each water company to be assessed in the most favorable way for it. This is another advantage of our proposal, since it does not demand excessive information for obtaining the global indicator.

This study, therefore, presents a pioneering and novel approach to assess the sustainability of water companies. To the best of our knowledge, there is neither any theoretical development nor empirical application that uses composite indicators to assess and/or compare the sustainability of water companies, for each dimension of sustainability and for all the dimensions, simultaneously. Thus, the dimensional composite indicators, in the first phase, allow evaluating the strengths and weaknesses of each water company in a particular dimension. The global indicator, in the second phase, provides a holistic performance perspective, and allows ranking the water companies. However, it provides information about the contribution of the different dimensions to the sustainability overall score.

In the next section, the methodology proposed is detailed. Section 3 introduces the case study, embracing 163 Portuguese water companies as well as the results obtained. Finally, the main conclusions derived from the research are presented in the last section.

2. A Two-Phase Evaluation Method

In this section, the methodology developed to construct the sustainability composite indicator is described, in order to evaluate the performance of water companies.

As previously mentioned, a two-phase procedure is proposed. In the first phase, following the proposal by [5], the composite indicator (sub-indicator) for each dimension of sustainability is calculated: $PSUI^d$ (Partial Sustainability Indicator of dimension d). In the second phase, these partial indicators form the basis from which the overall composite indicator is obtained, applying a variant of DEA named the "Benefit-of-the-Doubt" approach. Figure 1 shows the general scheme of the proposed approach.

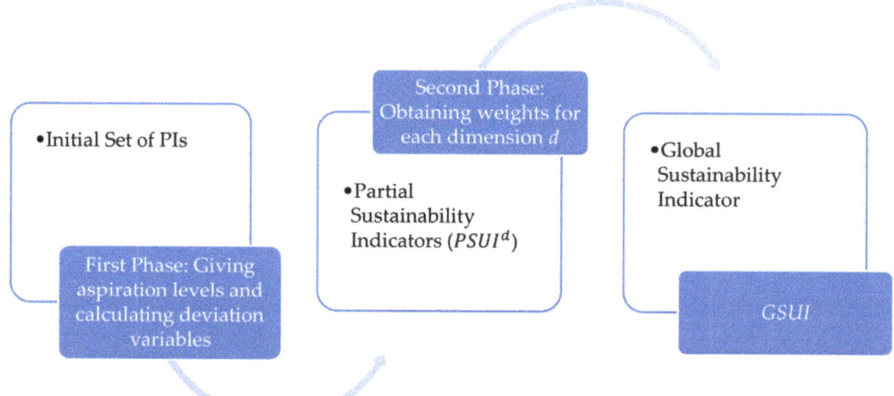

Figure 1. General scheme of the proposed approach.

Then, to calculate $PSUI^d$, $d = 1,2,\ldots,D$, it is necessary to distinguish between positive PIs (a larger value involves an improvement in the sustainability) and negative PIs (a larger value involves a decline in the sustainability). Let us suppose that the initial set of PIs is divided into D dimensions and there are M units to evaluate. For each $d \in D$, let us denote by J^d and K^d the number of positive and negative PIs, respectively, assigned to dimension d, and I^+_{i,j^d} and I^-_{i,k^d} the value of the i-th unit with respect to the j^d-th positive and k^d -th negative PI which belong to the dimension d ($i = 1,2,\ldots M$; $j^d = 1,2,\ldots J^d$; $k^d = 1,2,\ldots K^d$; $d = 1,2,\ldots D$).

Additionally, the performance of a unit is evaluated, regarding PIs, using the concept of "aspiration level", that is, the achievement level desired for the corresponding PI. Thus, it is possible to obtain a set of goals in line with the basic ideas underlying in GP approach [35]. Accordingly, let us assume that, for each positive PI, it is possible to give an aspiration level (denoted by $u^+_{j^d}$). It corresponds to the minimum value from which it is considered that a unit shows a suitable performance, regarding the aspect of sustainability evaluated by the PI. Thus, for the i-th unit, the goal corresponding to the j^d-th positive PI can be defined as follows:

$$I^+_{ij^d} + n^+_{ij^d} - p^+_{ij^d} = u^+_{j^d} \text{ with } n^+_{ij^d},\ p^+_{ij^d} \geq 0,\ \ n^+_{ij^d} \cdot p^+_{ij^d} = 0 \quad (1)$$

where $n^+_{ij^d}$, $p^+_{ij^d}$ represent the negative and positive deviation variables, respectively. Thus, if the goal is satisfied ($I^+_{ij^d} > u^+_{j^d}$), the negative deviation variable would be zero, and the positive deviation variable would measure the over-achievement of the goal (strength). Otherwise, if the goal is not satisfied ($I^+_{ij^d} < u^+_{j^d}$), the positive deviation variable would be zero and the negative deviation variable would quantify the under-achievement of the goal (weakness). It should be noted that, at least, one of the two deviation variables has to be zero. Consequently, for positive PIs, the negative deviation variables will be considered unwanted variables because a better-positioned company will achieve the aspiration level or a higher value.

In a similar way, for each negative PI, we have the following goal:

$$I^-_{ik^d} + n^-_{ik^d} - p^-_{ik^d} = u^-_{k^d} \text{ with } n^-_{ik^d},\ p^-_{ik^d} \geq 0,\ \ n^-_{ik^d} \cdot p^-_{ik^d} = 0 \quad (2)$$

Again, $n^-_{ik^d}$, $p^-_{ik^d}$ represent the negative and positive deviation variables, respectively. However, now, if the goal is satisfied ($I^-_{ik^d} < u^-_{k^d}$), the positive deviation variable would be zero and the negative deviation variable would quantify the under-achievement of the goal (strength). Otherwise, if the goal is not satisfied ($I^-_{ik^d} > u^-_{k^d}$), the negative deviation variable would be zero, and the positive deviation variable would quantify the over-achievement of the goal (weakness). Consequently, for negative PIs, the positive deviation variables will be considered unwanted variables because a better-positioned company will achieve the aspiration level or a lower value.

From all the above, at each dimension d, the strengths of each unit can be calculated by aggregating positive deviation variables, in case of positive PIs, and negative deviation variables, for the negative PIs. These variables are normalized by their corresponding aspiration levels to avoid the inadequate effects due to the use of different measurement scales of the initial set of PIs. Similarly, the weaknesses of each water company can be obtained as the sum of the normalized unwanted deviation variables (negative deviation for positive PIs and positive deviation for negative PIs divided by its corresponding aspiration level). Finally, the partial indicator for the i-th (i = 1,2, … M) unit, in the dimension d ($d \in D$) is determined by the difference between the strengths and weaknesses of this unit as follows:

$$\widetilde{PSUI}_i^d = \left(\sum_{j^d=1}^{J^d} \frac{p^+_{ij^d}}{u^+_{j^d}} + \sum_{k^d=1}^{K^d} \frac{n^-_{ik^d}}{u^-_{k^d}} \right) - \left(\sum_{j^d=1}^{J^d} \frac{n^+_{ij^d}}{u^+_{j^d}} + \sum_{k^d=1}^{K^d} \frac{p^-_{ik^d}}{u^-_{k^d}} \right) \tag{3}$$

Additionally, two fictitious units are introduced in the sample, representing the best and worst situation within the data base. For each positive indicator in dimension d, j^d, and negative PI, k^d, the value of the "best" unit (b) will be:

$$I^+_{bj^d} = Max_{i \in M}\left\{I^*_{ij^d}\right\},\ I^-_{bk^d} = Min_{i \in M}\left\{I^-_{ik^d}\right\} \tag{4}$$

and the value of the worst unit (w):

$$I^+_{wj^d} = Min_{i \in M}\left\{I^*_{ij^d}\right\},\ I^-_{wk^d} = Max_{i \in M}\left\{I^-_{ik^d}\right\} \tag{5}$$

For these fictitious units, their corresponding partial sustainability indicators are calculated. Finally, we can obtain the difference between $\widetilde{PSUI}_i^d$ with respect to the value reached by the worst unit and normalize this value by the difference between the partial sustainability indicator for the best and the worst unit, that is:

$$PSUI_i^d = \frac{\widetilde{PSUI}_i^d - \widetilde{PSUI}_w^d}{\widetilde{PSUI}_b^d - \widetilde{PSUI}_w^d},\quad i = 1,\ 2, \ldots,\ M;\ d \in D \tag{6}$$

The advantage of using $PSUI_i^d$ instead $\widetilde{PSUI}_i^d$ is that it offers a relative value between 0 and 1. In fact, it represents how far a unit is from the worst situation regarding the distance between the best and the worst situation. Additionally, this normalization does not distort the previously obtained results, but allows a more homogeneous and simple analysis of the dimensional results obtained.

Once the partial sustainability indicators for each dimension are obtained, the second phase consists of calculating the global sustainability indicator ($GSUI$). To do so, the "Benefit-of the-Doubt" approach [36], which is rooted DEA, is applied.

Now, for each unit a, $GSUI_a$ (a =1,2, … M) represents the weighted average of the partial indicators $PSUI_a^d$ ($d \in D$), which is obtained by solving the following optimization problem:

$$GSUI_a = Max \sum_{d \in D} w_a^d PSUI_a^d$$

Subject to:

$$\begin{aligned} &\sum_{d\in D} w_i^d PSUI_i^d \le 1 \;; i = 1,2,\ldots,a\ldots,\ldots M \\ L^d \le &\frac{w_i^d PSUI_i^d}{\sum_{d\in D} w_i^d PSUI_i^d} \le U^d \;; i = 1,2,\ldots,a\ldots,\ldots M; d \in D \\ &w_i^d \ge 0 \; i = 1,2,\ldots,a\ldots,\ldots M; d \in D \end{aligned} \tag{7}$$

where U^d and L^d are the upper and lower bounds allowed for the relative contribution of $PSUI^d$ to the global indicator. The aim of Equation (7) is to obtain the weights (assigned to the partial indicators) that maximize the global score ($GSUI_a$) for every unit a. Therefore, this model provides a relative objective performance value for each unit without requiring prior knowledge of the weights for the partial indicators [37]. These weights are endogenously determined solving Equation (7) and, by construction, $GSUI_a$ takes value between 0 (the worst situation) and 1 (the best situation).

In essence, Equation (7) is an output multiplier DEA model with multiple outputs (partial indicators) and a single "dummy input" with value equal to 1 for all the units [38]. In the DEA context, the contribution of each partial indicator to the value of the global indicator ($w_i^d PSUI_i^d$) is labelled as the "virtual output" of the corresponding dimension.

To avoid extreme situations, some constraints on the weights have been added to Equation (7). All partial indicators should have a relative contribution on the global indicator, that is, all the dimensions should be taken into account in the global score. For this reason, lower and upper bounds (U^d and L^d) have been established on the relative contribution of each partial indicator ($PSUI^d$).

Thus, the proposed approach offers a composite indicator which provides information about the contribution of each sustainability dimension to the global score. It allows to take into account the special characteristics of the units considered since the same importance does not need to be given to each dimension for the different units.

3. A Real Application

3.1. Data Description

Our aim is to use the concept of sustainability proposed by [20] to evaluate the performance of Portuguese water companies. In Portugal, we find two kinds of water companies: on the one hand, there are companies that provide services in all activities involved in the urban water cycle and, on the other hand, there are companies that focus on the distribution of drinking water and collection of wastewater. In any case, a national authority (ERSAR: Entidade Reguladora dos Serviços the Águas e Resíduos (www.ersar.pt)) regulates all companies. ERSAR states different regulatory functions over all the operators related to waste and water management. The statutes of ERSAR impose significant regulatory functions among the operators in charge of waste and water management in Portugal. Their concern is to respect customer rights and safeguard sustainability, as well as to provide economic visibility of the systems. In particular, this national authority applies the sunshine regulation model [39], which consists of sharing the information derived from a set of specific performance indicators that is provided by the operators. There are several factors that differentiate the Portuguese water companies, such as the management model or the regional location, among others. Portugal offers different management models for their water companies [19]: direct management (municipalities, municipalized services, and associations of municipalities); delegation (municipal-owned company or company established in partnership with the State (municipal or State-owned company), parishes, or user associations), and concession (municipal concessionaire or public–private partnership—municipality or municipalities and other private operators). In general, most of the municipalities receive the service directly from the municipal departments or municipal services with autonomy. This regulatory model has some strengths (the quality of service, the technical regulation and the access to information) but it also has some weaknesses (poor governance and failure to address identified problems). The evolution of the Portuguese water industry has been widely studied. However, some internal problems remain (water losses, poor staff productivity, . . .), in addition to the fact that the sector is excessively politicized. A more detailed description about this model of regulation and its characteristics can be found in [19].

To show the potential of the methodology proposed in the previous section, we consider an initial set of indicators applied to a set of Portuguese water companies. In the selection of these sustainability metrics, we take into account the availability of statistical data [37], as well as their relevance. The selection of these indicators is analogous to [12], whose data were obtained from the ERSAR list of Portuguese water companies in 2012. Nevertheless, on this occasion, data is updated to 2015 and, besides, the present work really makes use of the classification into three dimensions established in [12], in order to carry out the first phase of our approach. Then, 14 initial indicators are set, divided into three dimensions: social (5), environmental (5), and economic (4). In general, IS denote social indicators, whereas IEN are those related with environmental issues and IEC for the economic indicators.

Table 1 summarizes the main features of each initial PI, as well as the direction of improvement (negative or positive PIs), unit, average and standard deviation (for more details see Appendix A: Table A1). In particular, IS4, IS5, IEN4, and IEN5 are binary indicators, so they get a value of 1 if the water company has the certification, or 0 otherwise.

Table 1. Direction of improvement and statistical information from the initial set of PIs.

Acronym	Direction	Unit	Average	Standard Deviation
IS1	Positive	%	86.62	8.93
IS2	Positive	%	99.15	1.01
IS3	Positive	Days	1.46	0.9
IS4	Positive	-	0.15	0.36
IS5	Positive	-	0.09	0.29
IEN1	Negative	m^3/km/day	127.63	104.62
IEN2	Positive	%	0.56	2.69
IEN3	Negative	kWh (m^3/100 m)	0.88	0.65
IEN4	Positive	-	0.15	0.36
IEN5	Positive	-	0.31	0.47
IEC1	Negative	%	37.11	14.94
IEC2	Negative	Number/10^3 connections	2.15	1.05
IEC3	Positive	%	92.99	33.34
IEC4	Positive	-	49.16	26.31

Regarding water companies, Table 2 provides information related to the localization of them. In this sense, following the classification from Eurostat, the NUTS classification (Nomenclature of Territorial Units for Statistics) is a hierarchical system for dividing up the economic territory of the European Union. In particular, NUTS II are basic regions for the application of regional policies. Then, Portugal (continental) is divided into five regions or NUTS II (North, Centre, Metropolitan Area of Lisbon (MA Lisbon), Alentejo, and Algarve). Most of the water companies are located in the North (48) and Centre (58) regions, a large group is equally located in the Alentejo (30) region, and just a few of them are located in Algarve (12) and MA Lisbon (15) regions.

Table 2. Localization of the water companies and characteristics of the regions.

Region (NUTS II)	Water Companies	Area (Km^2)	Population (2011)	Pop/km^2	Share in National GDP % (2017)	GDP per Capita (€) (2017)
North	48	21,285	3,689,682	173.35	29.40%	16,000
Centre	58	28,217	2,327,755	82.49	18.90%	16,400
MA Lisbon	15	2802	2,821,876	1007.09	36.00%	24,700
Alentejo	30	27,292	757,302	27.75	6.50%	17,800
Algarve	12	4960	451,006	90.93	4.60%	20,500

Additionally, general information (obtained from Eurostat) about these NUTS II are shown in Table 2, in order to clarify the main characteristics of the regions in which the water companies are located. Thus, Centre and Alentejo regions are the largest areas, while MA Lisbon region is the smallest. Nevertheless, the last one presents the highest population per km^2 (1007.09 population). Finally, MA Lisbon region gets the highest GDP per capita (24,700 €), representing 36.00% of the total Portugal GDP; while the Centre region gets the lowest GDP per capita (16,000 €), representing the 29.40% of the total Portugal GDP.

3.2. Results and Discussion

Taking into account the case study described above, Figure 2 displays a visual scheme of our methodological approach to evaluate the case of the Portuguese water companies. Let us assume that every initial indicator is already assigned to a dimension. Observe that Phase 1 entails designing partial sustainable indicators ($PSUI^d$), in order to analyze the situation of water companies for each particular dimension, based on the information provided by the corresponding initial indicators. Afterwards, Phase 2 summarizes the information provided by these $PSUI^d$ into a global indicator ($GSUI$).

3.2.1. Phase 1: $PSUI_i^d$ Calculation

The first phase addresses the calculation of $PSUI_i^d$. To do this, the aspiration levels for each indicator have to be established. Our proposal follows previous works [5,12,40,41], so that, for positive initial indicators, the aspiration levels were set to the 80% value of the mean for each initial indicator; whereas in the case of negative initial indicators, the reciprocal percentage of the mean was used.

Results obtained are shown in Figure 3 (Water companies are listed following the ranking obtained for the global indicator in the second phase. The numbers associated with the water companies are provided in Table 5). For each water company, a set of three values is represented in different colors which denote each dimension. Note that, following the formulation of the dimensional indicators, the maximum value that a water company can get for each dimension is 1. Therefore, for the social dimension, it can be seen that five water companies obtain remarkable results in comparison to others: Águas de Cascais, EPAL, SMAS de Sintra, Vimágua and SMSB de Viana do Castelo. According to its location, Águas de Cascais, EPAL, and SMAS de Sintra are located in MA Lisbon region, and Vimágua and SMSB de Viana do Castelo in North region. It can be seen that most of the water companies obtain poor results for this dimension and that just a few of them obtain values greater than 0.5.

Likewise, an analysis within the environmental dimension reveals the good performance of Águas de Gondomar, Indaqua Matosinhos, Infraquinta, and Tavira Verde. Similarly, based on its location, Águas de Gondomar and Indaqua Matosinhos are located in the North region; and Infraquinta and Tavira Verde in the Algarve region. Aguas de Gondomar reaches the best position, since it provides the largest production of energy (IEN2). Note that each of these four better-positioned water companies produces between 14% and 20% of the energy that it uses. In general, most of the water companies obtain poor results in this dimension, too.

Additionally, in both of these dimensions, the values which indicate the certifications obtained for each water company plays an important role in the construction of the dimensional indicators, as determined in [12].

Finally, the results obtained for the economic dimension are ranged between 0.18 and 0.8 for all companies, highlighting Águas de Valongo and Indaqua Santo Tirso/Trofa, which are located in the North region. In particular, these water companies reach values greater than the average for all the initial indicators. In the data obtained, approximately the 56% of the water companies present an operating cost coverage ratio (IEC3) larger than the average.

In general, note that the best dimensional performance of the water companies is located in MA Lisbon, Algarve and North regions, despite the fact that more than a half of the companies (approximately 54%) are located in the other two regions (Alentejo and Centre regions).

In the same way, Table 3 shows the top 20 water companies for each dimension considered. It should be noted that there are four companies that appear among the top 20 in the three dimensions: Águas de Cascais, Águas de Valongo, Águas de Paredes, and Indaqua Matosinhos. Furthermore, 13 other companies stand out for two dimensions.

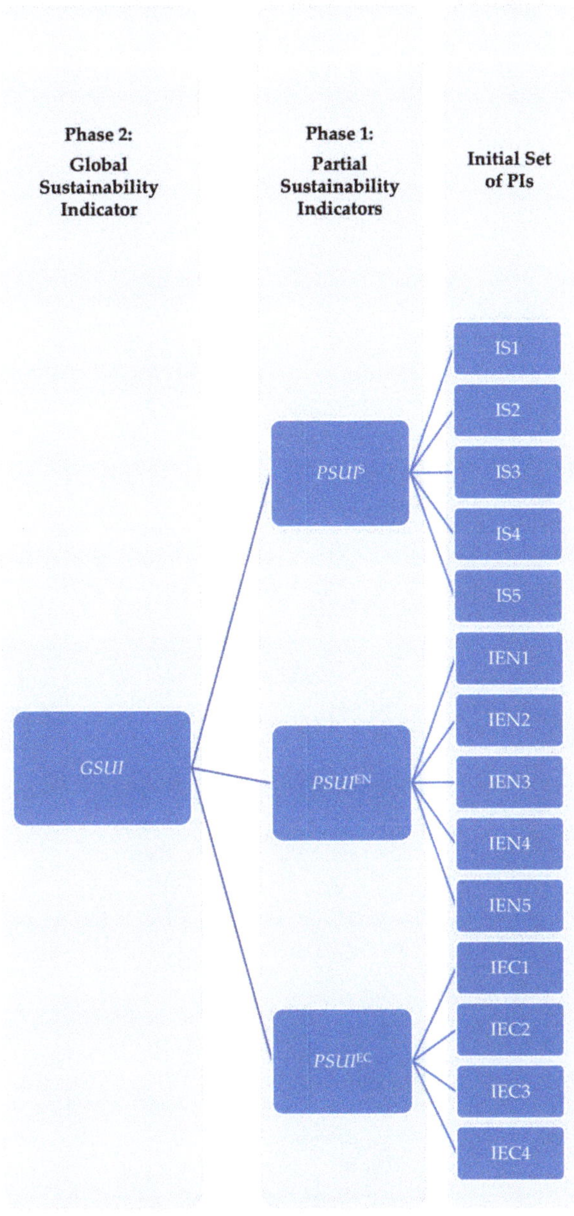

Figure 2. The two-phase approach applied to the case study.

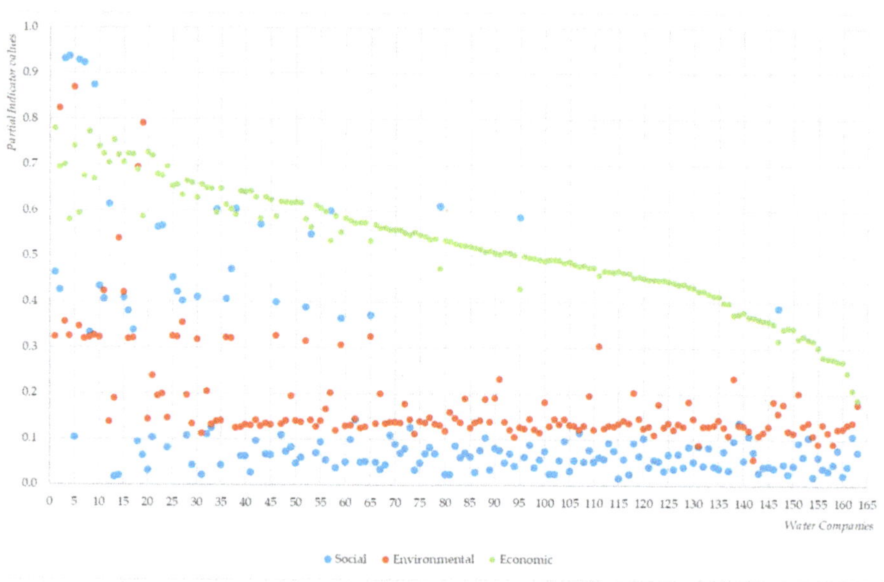

Figure 3. Dimensional results obtained for each of the 163 water companies.

Table 3. Top 20 water companies for each dimension.

	$PSUI_i^S$		$PSUI_i^{EN}$		$PSUI_i^{EC}$
4 VIMÁGUA	0.937	5 Águas de Gondomar	0.870	1 Águas de Valongo	0.779
3 Águas de Cascais	0.931	2 Indaqua Matosinhos	0.824	8 Indaqua Santo Tirso/Trofa	0.772
6 SMAS de Sintra	0.928	19 INFRAQUINTA	0.792	13 SM de Castelo Branco	0.754
7 EPAL	0.923	18 Tavira Verde	0.695	5 Águas de Gondomar	0.742
9 SMSB de Viana do Castelo	0.876	14 Águas de Barcelos	0.539	10 Águas de Paredes	0.740
12 Águas do Porto	0.615	11 Indaqua Feira	0.424	20 SMAS de Tomar	0.727
79 SMAS de Almada	0.610	15 Águas de Alenquer	0.422	11 Indaqua Feira	0.725
38 SMAS de Oeiras e Amadora	0.604	3 Águas de Cascais	0.356	16 INOVA	0.724
34 SMAS de Leiria	0.603	27 FAGAR - Faro	0.355	17 Indaqua Vila do Conde	0.724
163 CM de Miranda do Corvo	0.600	6 SMAS de Sintra	0.346	14 Águas de Barcelos	0.722
95 SM de Loures	0.586	9 SMSB de Viana do Castelo	0.327	21 Águas da Figueira	0.721
43 EMAS de Beja	0.570	46 Cartágua	0.326	15 Águas de Alenquer	0.707
23 Águas da Região de Aveiro	0.568	25 Águas de Mafra	0.326	12 Águas do Porto	0.705
22 Águas de Coimbra	0.564	4 VIMÁGUA	0.325	3 Águas de Cascais	0.700
53 CM de Santiago do Cacém	0.549	65 Aquamaior	0.325	24 Águas do Planalto	0.697
37 Aquaelvas	0.472	26 AGERE	0.324	2 Indaqua Matosinhos	0.695
1 Águas de Valongo	0.466	1 Águas de Valongo	0.323	18 Tavira Verde	0.689
25 Águas de Mafra	0.454	10 Águas de Paredes	0.323	22 Águas de Coimbra	0.679
10 Águas de Paredes	0.435	36 Águas de Ourém	0.323	23 Águas da Região de Aveiro	0.677
2 Indaqua Matosinhos	0.428	8 Indaqua Santo Tirso/Trofa	0.322	7 EPAL	0.676

Similarly, Table 4 shows the bottom 20 water companies for each dimension. In this case study, most of the companies (33) get bad results in just one of the three dimensions and only three water companies obtain poor results in the three dimensions: CM de Castelo de Paiva, CM de Arronches, and CM de Aljustrel.

Table 4. Bottom 20 water companies for each dimension.

	$PSUI_i^S$		$PSUI_i^{EN}$		$PSUI_i^{EC}$
137 CM de Avis	0.033	141 CM de Alijó	0.121	144 CM de Ferreira do Alentejo	0.361
124 CM de Mértola	0.032	93 CM de Armamar	0.121	145 CM de Marvão	0.359
157 CM de Castelo de Paiva	0.032	149 CM de Arronches	0.120	146 CM de Lousã	0.353
105 CM de Almodôvar	0.032	58 CM de Redondo	0.120	149 CM de Arronches	0.345
20 SMAS de Tomar	0.032	80 CM de Castro Verde	0.119	150 CM de Cabeceiras de Basto	0.344
41 INFRALOBO	0.028	144 CM de Ferreira do Alentejo	0.118	148 CM de Alfândega da Fé	0.343
143 CM de Pinhel	0.028	157 CM de Castelo de Paiva	0.116	152 CM de Murça	0.326
86 CM de Odemira	0.028	150 CM de Cabeceiras de Basto	0.116	151 CM de Sátão	0.321
149 CM de Arronches	0.025	99 CM de São Brás de Alportel	0.115	153 CM de Penalva do Castelo	0.320
117 CM de Caminha	0.025	31 CM de Póvoa de Varzim	0.113	154 CM de Aljustrel	0.317
102 CM de Évora	0.025	74 CM de Oliveira do Hospital	0.112	147 CM de Santa Marta de Penaguião	0.317
101 CM de Figueiró dos Vinhos	0.025	122 CM de Estremoz	0.112	155 CM de São João da Pesqueira	0.303
80 CM de Castro Verde	0.024	143 CM de Pinhel	0.110	156 CM de Castanheira de Pera	0.281
81 SMAS de Guarda	0.024	137 CM de Avis	0.110	157 CM de Castelo de Paiva	0.279
160 CM de Ourique	0.023	154 CM de Aljustrel	0.110	158 CM de Moimenta da Beira	0.277
31 CM de Póvoa de Varzim	0.022	94 CM de Vila Nova de Famalicão	0.105	159 CM de Sabrosa	0.274
14 Águas de Barcelos	0.020	155 CM de São João da Pesqueira	0.092	160 CM de Ourique	0.271
154 CM de Aljustrel	0.020	158 CM de Moimenta da Beira	0.091	161 CM de Tabuaço	0.247
13 SM de Castelo Branco	0.018	131 CM de Proença-a-Nova	0.088	162 CM de Penedono	0.209
115 CM de Grândola	0.016	142 INFRATROIA	0.057	163 CM de Miranda do Douro	0.187

In managerial terms, within the top-20 rankings, those companies that follow the municipal concessionaire management model obtain good results in the environmental (13) and economic dimensions (11). Additionally, in the social dimension, the ranking is leaded by a mixture of water companies following different management models: municipal or State-owned companies (VIMÁGUA and EPAL), municipal concessionaire (Águas de Cascais), or direct management (SMAS de Sintra and SMSB de Viana do Castelo).

The results obtained in this section are of great interest for water regulators. They enable the operators to learn from the best positioned water companies in each dimension and establish operative strategies in the correct direction with the aim of reducing the weaknesses in the mid-term. In general, social and environmental issues are still insufficiently integrated into management processes and there is room to improve these dimensions. Regulators should promote certification programs to encourage water companies to make necessary improvements in order to obtain these certifications.

3.2.2. Phase 2: $GSUI_a$ Calculation

Once the dimensional indicators ($PSUI_i^d$) are obtained for each water company, the optimization problem (Equation (7)) is applied in order to obtain the Global Sustainability Indicator proposed ($GSUI_a$). These solutions will provide the weights for social (S), environmental (EN), and economic (EC) dimension, maximizing the global score for each water company. In this way, this problem provides the weights for social, environmental, and economic dimension, maximizing the global score for each water company. The lower and upper bounds of the constraints are set to 0.001 and $+\infty$, respectively. This ensures that each dimension represents, at least, the 0.1% of the global score.

Figure 4 shows the values obtained for $GSUI_a$. It can be observed that there are five water companies (Águas de Cascais, Águas de Gondomar, Águas de Valongo, Indaqua Matosinhos, and Vimágua) that reach a value equal to 1 for the global indicator. EPAL, Indaqua Santo Tirso/Trofa and SMAS de Sintra are, also, very close to achieving a score of 1. It should be noted that 42.95% of the set of water companies obtain a global value greater than 0.70; in particular, 25.15% of water companies obtain a global value greater than 0.80. Then, a large group of water companies obtains good results, as they reach a value close to 1. Additionally, there are no companies obtaining a global value lower than 0.20.

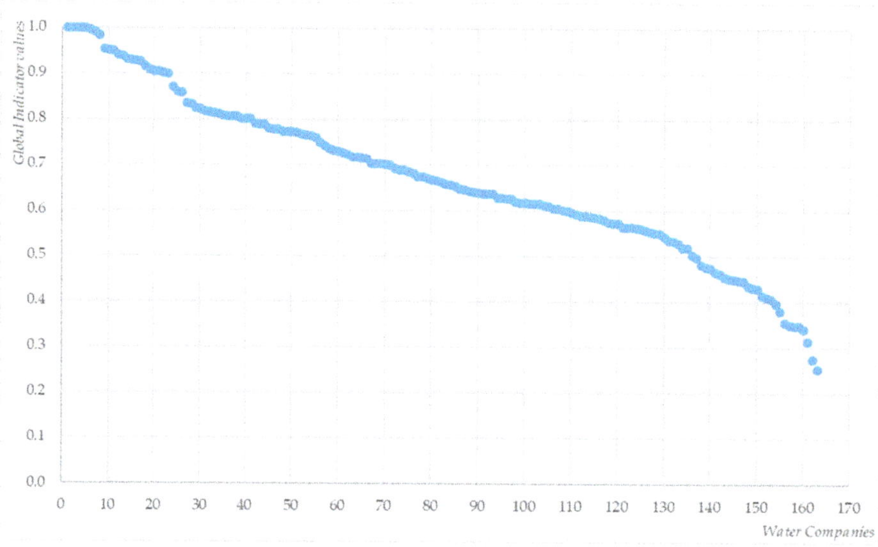

Figure 4. Global results obtained for each of the 163 water companies.

Table 5 lists all the water companies (163) according to the values obtained for the global indicator, $GSUI_a$. On the one hand, regarding the best results, four companies that belong to the top-20 for every dimension also appear within the best global indicator values. Moreover, between the companies ranked better, based on *GSUI*, there are 11 water companies whose *PSUI* value leads them to the top-20 in two different dimensions. On the other hand, there are three companies that were part of the bottom ranking for all the dimensions and, also, for the global indicator value. In particular, between the 20 worst positioned water companies, based on the global indicator *GSUI*, there are four water companies that also appeared among the worst results for two dimensions. The other 13 water companies included in this bottom 20 ranking obtained poor results in the economic dimension.

An analysis about the correlation between all the rankings is shown in Table 6 using a Kendall tau test. Correlations are significant at 1% level and values obtained show a high correlation between the economic dimension ranking and the global ranking (0.925). The rest of the correlations are similar and positives, and they are ranged between 0.26 and 0.40.

The proportional contribution differences can be explained by the particular profile characterizing each company. A global analysis reveals some influence of the location of the water companies in these lists. On the one hand, note that 12 of the top-20 water companies are located in the North region, three water companies are located in MA Lisbon, three water companies are located in the Centre region, and two water companies are located in the Algarve region. No one of the top-20 water companies is located in Alentejo region. Additionally, the geographical distribution of these water companies with the best performance on GPSUI might be grouped into two main locations along the Portuguese coast: those companies that are placed close to Oporto (for example: Águas de Valongo, Indaqua Matosinhos, Águas de Gondomar, Indaqua Santo Tirso/Trofa, Águas de Paredes, Indaqua Feira, Águas do Porto), and the ones close to the capital (Águas de Cascais, SMAS de Sintra, EPAL). On the other hand, within the bottom-20 water companies, none of them is located in the MA Lisbon or in Algarve region. Nevertheless, most of them are located in the North region (11), five water companies are located in the Alentejo region, and four water companies are located in the Centre region.

Table 5. Results for 163 water companies for global indicator.

Water Company	$GSUI_a$	Water Company	$GSUI_a$
1 Águas de Valongo	1.000	83 CM de Nisa	0.657
2 Indaqua Matosinhos	1.000	84 CM de Arganil	0.657
3 Águas de Cascais	1.000	85 CM de Porto de Mós	0.652
4 VIMÁGUA	1.000	86 CM de Odemira	0.648
5 Águas de Gondomar	1.000	87 Águas de Carrazeda	0.647
6 SMAS de Sintra	0.995	88 CM de Vale de Cambra	0.643
7 EPAL	0.989	89 CM de Arraiolos	0.640
8 Indaqua Santo Tirso/Trofa	0.983	90 CM de Espinho	0.638
9 SMSB de Viana do Castelo	0.953	91 SMAS de Vila Franca de Xira	0.636
10 Águas de Paredes	0.951	92 CM de Vila Viçosa	0.636
11 Indaqua Feira	0.949	93 CM de Armamar	0.636
12 Águas do Porto	0.940	94 CM de Vila Nova de Famalicão	0.629
13 SM de Castelo Branco	0.939	95 SM de Loures	0.627
14 Águas de Barcelos	0.931	96 CM de Ponte de Sor	0.627
15 Águas de Alenquer	0.930	97 CM de Ponte da Barca	0.626
16 INOVA	0.928	98 CM de Seia	0.620
17 Indaqua Vila do Conde	0.926	99 CM de São Brás de Alportel	0.617
18 Tavira Verde	0.915	100 CM de Se1bra	0.617
19 INFRAQUINTA	0.908	101 CM de Figueiró dos Vinhos	0.616
20 SMAS de Tomar	0.905	102 CM de Évora	0.615
21 Águas da Figueira	0.904	103 CM de Alandroal	0.615
22 Águas de Coimbra	0.901	104 CM de Aljezur	0.612
23 Águas da Região de Aveiro	0.899	105 CM de Almodôvar	0.610
24 Águas do Planalto	0.871	106 CM de Monção	0.605
25 Águas de Mafra	0.861	107 CM de Óbidos	0.604
26 AGERE	0.858	108 CM de Vila Nova de Foz Coa	0.601
27 FAGAR—Faro	0.836	109 CM de Arcos de Valdevez	0.599
28 CM de Albufeira	0.834	110 AMBIOLHÃO	0.596
29 CM de Moita	0.824	111 INFRAMOURA	0.592
30 Águas de S. João	0.823	112 CM de Montemor-o-Velho	0.588
31 CM de Póvoa de Varzim	0.816	113 CM de Cadaval	0.588
32 Luságua Alcanena—Gestão de Águas	0.815	114 CM de Terras de Bouro	0.586
33 EMAR de Portimão	0.813	115 CM de Grândola	0.585
34 SMAS de Leiria	0.811	116 CM de Alvaiázere	0.583
35 SMAS de Viseu	0.809	117 CM de Caminha	0.580
36 Águas de Ourém	0.807	118 CM de Bombarral	0.575
37 Aquaelvas	0.807	119 CM de Mora	0.573
38 SMAS de Oeiras e Amadora	0.806	120 CM de Alcoutim	0.572
39 Águas do Sado	0.801	121 CM de Nelas	0.565
40 Águas do Ribatejo	0.801	122 CM de Estremoz	0.564
41 INFRALOBO	0.801	123 CM de Mira	0.564
42 Águas da Azambuja	0.790	124 CM de Mértola	0.561
43 EMAS de Beja	0.789	125 SMAS de Peniche	0.560
44 SM de Alcobaça	0.788	126 CM de Lamego	0.556
45 CM de Marinha Grande	0.779	127 CM de Castro Daire	0.553
46 Cartágua	0.778	128 CM de Mourão	0.551
47 CM de Sines	0.777	129 CM de Penela	0.549
48 Águas do Lena	0.773	130 CM de Ponte de Lima	0.542
49 CM de Vila Verde	0.773	131 CM de Proença-a-Nova	0.535
50 CM de Seixal	0.772	132 CM de Soure	0.533
51 Penafiel Verde	0.770	133 CM de Chaves	0.529
52 Águas de Santarém	0.768	134 CM de Pedrógão Grande	0.520
53 CM de Santiago do Cacém	0.765	135 CM de Ferreira do Zêzere	0.519
54 SM de Nazaré	0.764	136 SMAS de Caldas da Rainha	0.503
55 Águas do Marco	0.759	137 CM de Avis	0.498
56 EMAR de Vila Real	0.748	138 CM de Vimioso	0.482
57 CM de Miranda do Corvo	0.742	139 CM de Vila de Rei	0.476
58 CM de Redondo	0.734	140 CM de Vila Nova de Poiares	0.475
59 Aquafundalia	0.731	141 CM de Alijó	0.466
60 CM de Reguengos de Monsaraz	0.730	142 INFRATROIA	0.461
61 CM de Sousel	0.726	143 CM de Pinhel	0.455
62 CM de Mogadouro	0.722	144 CM de Ferreira do Alentejo	0.451
63 CM de Mealhada	0.718	145 CM de Marvão	0.450
64 CM de Almeida	0.718	146 CM de Lousã	0.447

Table 5. *Cont.*

Water Company	$GSUI_a$	Water Company	$GSUI_a$
65 Aquamaior	0.715	147 CM de Santa Marta de Penaguião	0.446
66 CM de Mangualde	0.713	148 CM de Alfândega da Fé	0.435
67 SMAS de Montijo	0.703	149 CM de Arronches	0.432
68 CM de Barreiro	0.703	150 CM de Cabeceiras de Basto	0.431
69 CM de Pombal	0.703	151 CM de Sátão	0.415
70 SMAS de Torres Vedras	0.701	152 CM de Murça	0.410
71 CM de Penacova	0.699	153 CM de Penalva do Castelo	0.406
72 CM de Lagos	0.692	154 CM de Aljustrel	0.396
73 CM de Góis	0.689	155 CM de São João da Pesqueira	0.381
74 CM de Oliveira do Hospital	0.689	156 CM de Castanheira de Pera	0.355
75 CM de Montemor-o-Novo	0.684	157 CM de Castelo de Paiva	0.350
76 CM de Palmela	0.682	158 CM de Moimenta da Beira	0.347
77 CM de Ansião	0.675	159 CM de Sabrosa	0.347
78 SM de Abrantes	0.674	160 CM de Ourique	0.341
79 SMAS de Almada	0.671	161CM de Tabuaço	0.314
80 CM de Castro Verde	0.666	162 CM de Penedono	0.275
81 SMAS de Guarda	0.665	163 CM de Miranda do Douro	0.253
82 CM de Melgaço	0.662		

Table 6. Kendall tau test.

Rank Correlation Coefficient	Global Ranking	Social Ranking	Environmental Ranking	Economic Ranking
Global Ranking	1.000			
Social Ranking	0.324 **	1.000		
Environmental Ranking	0.395 **	0.369 **	1.000	
Economic Ranking	0.925 **	0.261 **	0.361 **	1.000

**: significance level at 1%.

In relation to the weights obtained by solving Equation (7), Table 7 summarizes the maximum and minimum values for the virtual outputs obtained, as well as their mean and standard deviation for each dimension. The economic dimension, in general, is the one with the largest virtual outputs in the global aggregation. As the reader may observe in Figure 3, almost all water companies obtain similar (good) values in this dimension. On the contrary, the social and the environmental dimensions lose importance in the global weighting. The results show how some companies obtained very good results in these dimensions but, at the same time, a larger proportion of the companies obtained relatively poor results.

Table 7. Statistical information for weights calculated.

Descriptive Statistics	Social	Environmental	Economic
Minimum	0.001	0.001	0.003
Maximum	0.991	0.990	0.966
Mean	0.045	0.029	0.598
Standard Deviation	0.156	0.127	0.197

Figure 5 shows the percentage contribution of each partial indicator to the value of the global composite indicator, for the top-20 water companies. Differences in percentage contribution can be explained by the particular profile characterizing each company. Despite the good performance of those 11 companies within the top-20 at each dimension, in the global score there is no company that displays a balanced contribution among the three dimensions. Then, if a balance between dimensions is searched, the best water companies of the global indicator are not a good reference for the others, in this context. In general, water companies should seek to improve their results in the dimension in which they obtained the worst results in the first phase, without neglecting the maintenance of good performance in those dimensions in which they obtained good results. Moreover, focusing on the companies that have a value equal to 1 in the global indicator, Águas de Cascais and VIMÁGUA exhibit a greater contribution of the social dimension, while Indaqua Matosinhos and Águas de Gondomar prioritize environmental dimension. However, Águas de Valongo stands out for the percentage contribution of the economic dimension, while a minor importance of the others. In the rest of companies, among those with a value of the global indicator close to 1, it is worth mentioning that Indaqua Feira and Águas de Alenquer present a similar performance with an analogous contribution of each dimension, around 8% for the social dimension, 9% for the environmental dimension, and 83% for the economic one. In general, a trade-off between the economic dimension and the others can be observed jointly.

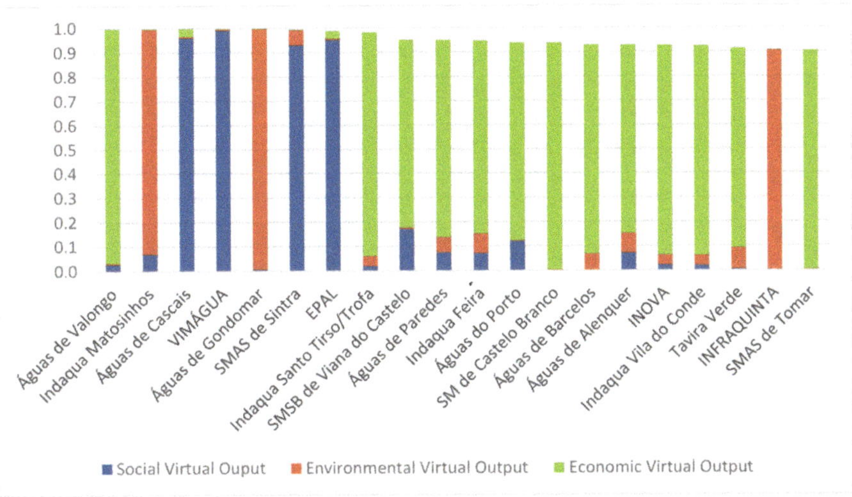

Figure 5. Contribution of each dimension partial indicator to the value of global composite indicator.

In managerial terms, within the top-20 ranking, ten companies that follows municipal concessionaire management model obtain good results the global indicator. The other ten companies present a fair distribution between the other two management models. Additionally, note that all water companies belonging to the bottom-20, use the municipal direct management model.

Then, the proposed approach allows evaluating the strengths and weaknesses of each water company in a particular dimension and, at the same time, provides information about the contribution of the dimensions to the sustainability overall score. These are the main advantages in comparison to the previous methodology. In fact, if the initial indicators are directly aggregated using the methodology based on goal programming, the results obtained regarding the best and the worst global performance of the water companies are similar to those obtained with the proposed two-phase approach. In particular, the top-13 using the previous methodology is formed by water companies that appear in our global top-20. Nevertheless, using the previous methodology, the advantages named above disappear because dimensional results are not obtained and, therefore, the contribution of them to the global indicator is missing.

Consequently, in light of the results obtained, it is necessary to perform some transformations towards sustainability with a balanced percentage contribution of each dimension. Implementing appropriate programs that highlight social and environmental aspects is required to address global sustainability in an adequate manner. Nevertheless, the proposed approach allows a better observation of the differences among the water companies, dimensional and globally. It eases identifying strengths and weaknesses of the companies, helping the decision-maker to set strategies to improve the medium- and long-term sustainability of such companies.

4. Conclusions

Despite the multiple benefits brought by following an efficient performance, in water management only a few works provide alternatives. In this context, benchmarking plays an important role. Normally, to study the efficiency requires information collected by indicators. However, some difficulties arise when dealing with several indicators and their interpretation. In order to overcome these problems, CIs are introduced in this field, providing different strategies to aggregate the indicators into a unique score that summarizes the information.

In particular, this work provides an alternative methodology to evaluate the performance of a set of Portuguese water companies following two steps. First, in order to analyse the economic, social and environmental dimensions, we divide the initial set of indicators into these three dimensions and construct a partial sustainable index for each of them, inspired by GP. In general, water companies present the largest value for the economic partial sustainability indicator (PSUIEC), whereas the partial sustainable index for the social and environmental dimensions present poor scores, the former being slightly lower. In particular, ranking these results, we find that just a few water companies stand out among the top 20 best scores in the three dimensions, simultaneously. This fact could be translated into policies to improve social and environmental aspects of the water companies. The second step uses a variant of DEA to provide a global performance index that uses the information provided by the partial indicators for each company. As a result, a large percentage of water companies obtain a global score over 0.7, whereas no companies show a value below 0.2. However, an individual analysis of the contribution of each dimension to the global score shows no equilibrium.

Furthermore, in this analysis one may observe two profiles: on the one hand, many water companies present a good global score due to the value they achieve in the economic partial sustainability indicator, whereas, on the other hand, the good global results of the other companies are due to their performance in the social and environmental dimensions, jointly. The results obtained show that the water companies, in the Portuguese context, do not manage their activity in a balanced way from the social, environmental, and economic point of view. Consequently, there are no water companies, in this context, that can be considered a "good benchmark" for the rest, so that they achieved good results in the three dimensions in a balanced way. In this case, each water company should seek to improve their results in the dimension(s) with lower contribution in the sustainability, taking into account the scores in the first phase, without neglecting the maintenance of good performance in those dimensions in which good results are obtained. In this context, as a future line of research, it would be interesting to define an ideal company that reaches a good percentage contribution of each dimension on the sustainability, and then compare the sample of water companies with this one ideal.

Nevertheless, this work introduces an alternative to assess the sustainability of water companies in two phases. It permits assessing and/or comparing the dimensional sustainability in the first phase, and to provide a holistic performance perspective in the second phase, generating a ranking of the water companies. The proposed approach could be very useful for water regulators: (a) to verify the effectiveness of existing policies; (b) to support decision making in concrete dimensions; and (c) to monitor global trends. In other words, measuring sustainability, holistically and for dimension, will allow water regulators to make critical decisions and, if needed, implement corrective measures to improve it and do it in the correct direction.

Author Contributions: All authors contributed to the paper as follows. Conceptualization, F.P., L.D.-A. and T.G.; Methodology, F.P., L.D.-A. and T.G.; Software, F.P.; Validation, F.P., L.D.-A. and T.G.; Formal Analysis, F.P., L.D.-A. and T.G.; Investigation, L.D.-A.; Resources, F.P., L.D.-A. and T.G.; Data Curation T.G.; Writing—Original Draft Preparation, F.P., L.D.-A. and T.G.; Writing—Review and Editing, F.P., L.D.-A. and T.G.; Visualization, F.P., L.D.-A. and T.G.; Supervision, F.P., L.D.-A. and T.G.

Funding: This research was funded by Ministerio de Ciencia, Innovación y Universidades from Spain, grant number ECO2016-76567-C4-4-R; by the Consejería de Conocimiento, Investigación y Universidad from Junta de Andalucia [SEJ417] and by Universidad de Málaga, grant number PPIT.UMA.B1.2017.21.

Conflicts of Interest: The authors declare no conflict of interest.

Appendix A

Table A1. Initial set of PIs.

Dimension	Acronym	Performance Indicator
Social	IS1	Service coverage (% of the households for which the water company provides effective connected service)
	IS2	Drinking water quality safety (% of water supplied that meets the legal quality requirements)
	IS3	Reserve capacity for treated water (capacity to supply water of the water company if new water resources are not available)
	IS4	Certification of management systems for occupational risk and health issues at work
	IS5	Other certifications (corporate social responsibility, consumer protection mechanisms, …)
Environmental	IEN1	Water losses in the network (volume of drinking water lost/km/day)
	IEN2	Internal power generation (% of energy used own-generated by the water company)
	IEN3	Energy efficiency in pumping water (average consumption of energy for water pumping)
	IEN4	Certification of management systems (environmental responsibility, environmental impact assessment mechanisms …)
	IEN5	Certification of management systems for water quality issues
Economic	IEC1	Non-revenue water (% of water that is supplied but not invoiced)
	IEC2	Adequacy of staffing (number of full time equivalent employed / 1000 water supply connections)
	IEC3	Operating cost coverage ratio (total annual operational revenues / total annual operational costs)
	IEC4	Index of knowledge about infrastructure and asset management

References

1. Carvalho, P.; Marques, R.C.; Berg, S. A meta-regression analysis of benchmarking studies on water utilities market structure. *Util. Policy* **2012**, *21*, 40–49. [CrossRef]
2. Molinos-Senante, M.; Gómez, T.; Garrido-Baserba, M.; Caballero, R.; Sala-Garrido, R. Assessing the sustainability of small wastewater treatment systems: A composite indicator approach. *Sci. Total Environ.* **2014**, *54*, 607–617. [CrossRef] [PubMed]
3. Giannetti, B.F.; Bonilla, S.H.; Silva, C.C.; Almeida, C.M.V.B. The reliability of experts' opinions in constructing a composite environmental index: The case of ESI 2005. *J. Environ. Manag.* **2009**, *90*, 2248–2459. [CrossRef] [PubMed]
4. Voces, R.; Díaz-Balteiro, L.; Romero, C. Characterization and explaination of the sustainability of the European wood manufacturing industries: A quantitative approach. *Expert Syst. Appl.* **2012**, *39*, 6618–6627. [CrossRef]
5. Blancas, F.J.; Caballero, R.; González, M.; Lozano-Oyola, M.; Pérez, F. Goal programming synthetic indicators: An application for sustainable tourism in Andalusian coastal countries. *Ecol. Econ.* **2010**, *69*, 2158–2172. [CrossRef]
6. Pérez, V.; Guerrero, F.; González, M.; Pérez, F.; Caballero, R. Composite indicator for the assessment of sustainability: The case of Cuban nature-based tourism destinations. *Ecol. Indic.* **2013**, *29*, 316–324. [CrossRef]
7. Diaz-Balteiro, L.; Voces, R.; Romero, C. Making sustainability rankings using compromise programming. An application to European paper industry. *Silv. Fenn.* **2011**, *45*, 761–773. [CrossRef]
8. Sayed, H.; Hamed, R.; Hosny, S.H.; Abdelhamid, A.H. Avoiding Ranking Contradictions in Human Development Index Using Goal Programming. *Soc. Indic. Res.* **2018**, *138*, 405–442. [CrossRef]
9. Guijarro, F.; Poyatos, J.A. Designing a Sustainable Development Goal Index through a Goal Programming Model: The Case of EU-28 Countries. *Sustainability* **2018**, *10*, 3167. [CrossRef]
10. Valcárcel-Aguiar, B.; Murias, P. Evaluation and Management of Urban Liveability: A Goal Programming Based Composite Indicator. *Soc. Indic. Res.* **2019**, *142*, 689–712. [CrossRef]
11. Xavier, A.; Costa Freitas, M.B.; Fragoso, R.; Rosário, M.S. A regional composite indicator for analysing agricultural sustainability in Portugal: A goal programming approach. *Ecol. Indic.* **2018**, *89*, 84–100. [CrossRef]
12. Molinos-Senante, M.; Marques, R.C.; Pérez, F.; Gómez, T.; Sala-Garrido, R.; Caballero, R. Assessing the Sustainability of water companies: A synthetic indicator approach. *Ecol. Indic.* **2016**, *61*, 577–587. [CrossRef]

13. Cherchye, L.; Moesen, W.; Rogge, N.; Van Puyenbroeck, T. *Creating Composite Indicators with DEA Analysis: The Case of the Technology Achievement Index*; Joint Research Centre, European Commission: Ispra, Italy, 2006.
14. Murias, P.; de Miguel, J.C.; Rodríguez, D. A composite indicator for university quality assessment: The case of Spanish higher education system. *Soc. Indic. Res.* **2008**, *89*, 129–146. [CrossRef]
15. Castellet, L.; Molinos-Senante, M. Efficiency assessment of wastewater treatment plants: A data envelopment analysis approach integrating technical, economic, and environmental issues. *J. Environ. Manag.* **2016**, *167*, 160–166. [CrossRef]
16. Hernández-Sancho, F.; Sala-Garrido, R. Technical efficiency and cost analysis in wastewater treatment processes: A DEA approach. *Desalination* **2009**, *249*, 230–234. [CrossRef]
17. Dong, X.; Zhang, X.; Zeng, S. Measuring and explaining eco-efficiencies of wastewater treatment plants in China: An uncertainty analysis perspective. *Water Res.* **2017**, *112*, 195–207. [CrossRef] [PubMed]
18. Holden, E.; Linnerud, K.; Banister, D. Sustainable development: Our Common Future revisited. *Glob. Environ. Chang.* **2014**, *26*, 130–139. [CrossRef]
19. Pinto, F.S.; Costa, A.S.; Figueira, J.R.; Marques, R.C. The quality of service: An overall performance assessment for water utilities. *Omega* **2017**, *69*, 115–125. [CrossRef]
20. *WECD, Our Common Future*; Oxford University Press: Oxford, UK, 1987.
21. Lo Storto, C. Efficiency, conflicting goals and trade-offs: A nonparametric analysis of the water and wastewater service industry in Italy. *Sustainability* **2018**, *10*, 919. [CrossRef]
22. Arnold, M. The lack of strategic sustainability orientation in German water companies. *Ecol. Econ.* **2015**, *117*, 39–52. [CrossRef]
23. Moller, K.A.; Fryd, O.; De Neergaard, A.; Magid, G. Economic, environmental and socio-cultural sustainability of three constructed wetlands in Thailand. *Environ. Urban.* **2012**, *24*, 305–323. [CrossRef]
24. Marques, R.C.; da Cruz, N.F.; Pires, J.S. Measuring the sustainability of urban water services. *Environ. Sci. Policy* **2015**, *54*, 142–151. [CrossRef]
25. Aydin, N.Y.; Mays, L.; Schmitt, T. Sustainability assessment of urban water distribution systems. *Water Res. Manag.* **2014**, *28*, 4373–4384. [CrossRef]
26. Hamouda, M.A.; Nour El-Din, M.M.; Moursy, F.I. Vulnerability assessment of water resoiurces systems in the Eastern Nile basin. *Water Res. Manag.* **2009**, *23*, 2697–2725. [CrossRef]
27. Lundie, S.; Peters, G.; Ashbolt, N.; Lai, E.; Livingston, D. A sustainability framework for the Australian water industry. *Water* **2010**, *33*, 83–88.
28. Schulz, M.; Short, M.D.; Peters, G.M. A streamlined sustainability assessment tool for improved decision making in the urban water industry. *Integr. Environ. Assess. Manag.* **2012**, *8*, 183–193. [CrossRef] [PubMed]
29. Ahn, J.; Kang, D. Optimal planning of water supply system for long-term sustainability. *J. Hydro Environ. Res.* **2014**, *8*, 410–420. [CrossRef]
30. Duarte, A.A.L.S.; Rodrigues, G.M.C.; Ramos, R.A.R. A global service quality index to evaluate the performance and sustainability in water supply utilities. *WSEAS Trans. Environ. Dev.* **2009**, *5*, 759–769.
31. Bana-e-Costa, C.A.; Vansnick, J.C. MACBETH-An interactive path towards the construction of cardinal value functions. *Int. Trans. Oper. Res.* **1994**, *1*, 489–500. [CrossRef]
32. Almeida-Dias, J.; Figueira, J.R.; Roy, B. ELECTRE TRI-C: A multiple-criteria sorting method based on characteristic reference actions. *Eur. J. Oper. Res.* **2010**, *240*, 565–580. [CrossRef]
33. Blancas, F.J.; Caballero, R.; Lozano-Oyola, M.; Pérez, F. The assessment of sustainable tourism: Application to Spanish coastal destinations. *Ecol. Indic.* **2010**, *10*, 484–492. [CrossRef]
34. Environmental Protection Agency. *Planning for Sustainability. A Handbook for Water and Wastewater Utilities*; EPA: Seattle, DC, USA, 2012.
35. Diaz-Balteiro, L.; González-Pachón, J.; Romero, C. Measuring systems sustainability with multi-criteria methods: A critical review. *Eur. J. Oper. Res.* **2017**, *258*, 607–616. [CrossRef]
36. Cherchye, L.; Moesen, W.; Rogge, N.; van Puyenbroeck, T. An introduction to "benefit of the Doubt" Composite Indicators. *Soc. Indic. Res.* **2006**, *82*, 111–145. [CrossRef]
37. Zhou, P.; Ang, B.W.; Poh, K.L. A mathematical programming approach to constructing. *Ecol. Econ.* **2007**, *62*, 291–297. [CrossRef]
38. Despotis, D.K. Measuring human development via data envelopment analysis: The case of Asia and the Pacific. *Omega* **2005**, *33*, 385–390. [CrossRef]

39. Marques, R.C.; Simões, P. Does the sunshine regulatory approach work? Governance and regulation model of the urban waste services in Portugal. *Resour. Conserv. Recycl.* **2008**, *52*, 1040–1049. [CrossRef]
40. Lozano-Oyola, M.; Blancas, F.J.; González, M.; Caballero, R. Sustainable tourism indicators as planning tools in cultural destinations. *Ecol. Indic.* **2012**, *18*, 659–675. [CrossRef]
41. Pérez, F.; Molinos-Senante, M.; Gómez, T.; Caballero, R.; Sala-Garrido, R. Dynamic goal programming synthetic indicator: An application for water companies sustainability assessment. *Urb. Water J.* **2018**, *15*, 592–600. [CrossRef]

Article

Assessment of Energy Efficiency and Its Determinants for Drinking Water Treatment Plants Using A Double-Bootstrap Approach

María Molinos-Senante [1,2,*] and Ramón Sala-Garrido [3]

1 Departamento de Ingeniería Hidráulica y Ambiental, Pontificia Universidad Católica de Chile, Av. Vicuña Mackenna 4860, Santiago, Chile
2 Centro de Desarrollo Urbano Sustentable CONICYT/FONDAP/15110020, Av. Vicuña Mackenna 4860, Santiago, Chile
3 Departamento de Matemáticas para la Economía y la Empresa, Universidad de Valencia, Avd. Tarongers S/N, 46023 Valencia, Spain; sala@uv.es
* Correspondence: mmolinos@uc.cl; Tel.: +56-223-544-219

Received: 21 January 2019; Accepted: 19 February 2019; Published: 25 February 2019

Abstract: To achieve energy and climate goals, the energy performance of current and future drinking water treatment plants (DWTPs) must be improved. A few studies have evaluated the energy efficiency of these facilities using data envelopment analysis (DEA), however, they have ignored the deterministic nature of the DEA method. To overcome this limitation, a double-bootstrap DEA approach was used in this study to estimate the energy efficiency of DWTPs. For a sample of Chilean DWTPs, bias-corrected energy efficiency scores were computed with consideration of data variability, and the determinants of DWTP energy efficiency were explored. Most DWTPs in the sample had much room for the improvement of energy efficiency. In the second stage of analysis, facility age, the volume of water treated, and the technology used for treatment were found to influence DWTP energy efficiency. These findings demonstrate the importance of using a reliable and robust method to evaluate the energy efficiency of DWTPs, which is essential to support decision making and to benchmark these facilities' energy performance.

Keywords: data envelopment analysis; energy efficiency; performance; bootstrap; water treatment

---

1. Introduction

In the context of climate change, energy demand for urban water supplies has emerged as a relevant issue [1]. In the future, more energy is expected to be required to treat and supply drinking water to citizens due to the adaption of water systems to the effects of climate change and to new regulatory requirements. Goal 6 of the Sustainable Development Goals adopted by the United Nations (UN) in 2015 involves the achievement of universal and equitable access to safe and affordable drinking water for all by 2030. The achievement of this goal will require the construction of many more drinking water treatment plants (DWTPs), which will increase the amount of energy required for drinking water supplies worldwide.

Urban water utilities use energy to extract, convey, treat, and distribute drinking water. Previous studies have focused on the evaluation of energy requirements for one or several activities in the urban water cycle. Several studies have focused on the quantification and evaluation of the economic and environmental effects of energy use to supply drinking water to major cities and on the comparison of energy use among cities and countries [2,3]. Other studies have involved a more detailed analysis of individual drinking-water supply stages and quantification of the energy required to treat raw water, i.e., the energy used by DWTPs [4]. The aim of these studies was to quantify and compare the energy intensity of DWTPs, that is, the energy consumed (kWh) per unit volume (m^3) of drinking water produced (kWh/m^3) [5]. However, as Santana et al. [6] and Molinos-Senante and Sala-Garrido [7] determined, the energy required to treat raw water depends on several factors, including the quality of the raw water and of drinking water standards, as well as the water treatment technology used. Sowby and Burian [8] also emphasized this issue; after analyzing the energy requirements for drinking water supplies in 109 cities in the United States, they concluded that energy intensity is an overly simplistic metric that is not adequate for the comparison of DWTP performance.

To overcome the comparability limitation and to facilitate benchmarking between DWTPs, a few studies have focused on evaluating DWTP energy efficiency. Molinos-Senante and Sala-Garrido [9] defined energy efficiency as a "synthetic index that incorporates both the quality of the raw water being processed and the energy required to treat it." Although several methods can be used to estimate the energy efficiency of DWTPs, data envelopment analysis (DEA) has been used in the few studies conducted to date. DEA is a non-parametric method based on mathematical programming techniques that integrates multiple inputs (energy use) and outputs (concentrations of several pollutants removed from raw water and the volume of treated water) into a synthetic index; namely, the energy efficiency score. Molinos-Senante and Guzman [10] computed the energy efficiency of a sample of DWTPs using the DEA approach, investigating the presence of economies of scale in these facilities. Applying the more advanced metafrontier DEA model, Molinos-Senante and Sala-Garrido [9] compared energy efficiency among DWTPs using different treatment technologies. Recently, Ananda [11,12] computed the environmental efficiency and productivity change of a sample of 49 Australian urban water utilities using DEA, with a focus on economic issues, but integration of greenhouse gas emissions as undesirable outputs.

These studies [9–12] contributed to the literature by providing estimated energy efficiency scores for DWTPs derived from the application of a holistic and integrated approach. However, they have ignored the deterministic nature of the DEA methodology; as statistical inferences cannot be drawn from conventional DEA (energy) efficiency scores [11] and regression analysis cannot be conducted to explore the determinants of previously estimated scores [12]. Moreover, conventional DEA models do not integrate data variability into the (energy) efficiency assessment, which negatively impacts the robustness and reliability of the results.

In the framework of efficiency assessment, two main alternative methodological approaches have been proposed to explore the causality between factors and efficiency scores. Cazals et al. [13] proposed the $order - m$ method, in which a fraction of the sample is used to estimate efficiency scores. However, selection of the m value is challenging, and it affects efficiency scores [14]. Alternatively, Simar and Wilson [15] proposed a double-bootstrap DEA procedure for the estimation of efficiency scores that overcomes the two main limitations of conventional DEA models; i.e., it allows exploration of the determinants or factors affecting efficiency scores [16], and it permits bias correction and the calculation of confidence intervals for the scores. Despite the relevance of this type of analysis, the bootstrap approach has not been used to evaluate the energy efficiency of DWTPs.

Against this background, the objectives of this study were twofold. The first objective was to assess the energy efficiency of a sample of DWTPs with consideration of data variability, i.e., to estimate bias-corrected energy efficiency scores and their confidence intervals. The second objective was to explore the determinants DWTP energy efficiency. To do so, we employed the double-bootstrap DEA approach proposed by Simar and Wilson [17]. Empirical application was performed with a large sample (n = 146) of Chilean DWTPs. Although many scholars have examined the energy intensity of urban water cycle activities in recent times, few studies have assessed the energy efficiency of water treatment plants and none has involved the application of a robust methodological approach such as double-bootstrap DEA. This paper contributes to the current body of literature in the water–energy nexus field by presenting for the first time bias-corrected energy efficiency scores and discussing factors affecting the energy efficiency of a sample of DWTPs.

2. Methods

2.1. Energy Efficiency Estimation

Conventional DEA models, such as those proposed by Charnes et al. [18] and Banker et al. [19] (the CCR (Charnes, Cooper and Rhodes) and BCC (Banker, Charnes and Cooper), respectively) and subsequent developments, have been employed widely to evaluate the efficiency of water treatment facilities, such as WWTPs and DWTPs. DEA models can be used with a constant returns to scale (CRS) or variable returns to scale (VRS) technique. With the CRS approach, outputs increase in proportion to inputs, and producers (DWTPs in this case) are assumed to be able to linearly scale inputs and outputs without changing efficiency. By contrast, with the VRS approach, an increase or decrease in input or output does not result in a proportional change in outputs or inputs, respectively. Previous studies [1,5] have shown that the energy consumed by DWTPs to produce drinking water does not depend linearly on the pollutants removed from raw water. Hence, in this study, and as in Molinos-Senante and Sala-Garrido's studies [9], a VRS DEA model was employed to evaluate DWTP energy efficiency.

Moreover, DEA models can have an input or output orientation, depending on whether the aim of the units analyzed (DWTPs) is to minimize the use of inputs or to maximize the production of outputs. In this case study, an input orientation was adopted because the main objective of DWTPs is to produce drinking water that complies with the legal quality standards using minimum energy. The quantity of pollutants to be removed depends on the quality of the raw water and the thresholds defined by drinking water standards, which are external to the water utilities.

Let us assume that we have a set of DWTPs, $j = 1,2\ldots,N$, each using a vector of M inputs, $x_j = (x_{1j}, x_{2j}, \ldots, x_{Mj})$, to produce a vector of S outputs, $y_j = (y_{1j}, y_{2j}, \ldots, y_{Sj})$. Assuming VRS technology, the input-oriented DEA model is denoted as follows:

$$
\begin{array}{ll}
Min\ \theta_j & \\
s.t. & \\
\sum_{k=1}^{N} \lambda_k x_{ik} \le \theta_j x_{ij} & 1 \le i \le M \\
\sum_{k=1}^{N} \lambda_k y_{rk} \ge y_{rj} & 1 \le r \le S \\
\sum_{k=1}^{N} \lambda_k = 1 & \\
\lambda_k \ge 0 & 1 \le k \le N,
\end{array}
\tag{1}
$$

where θ_j is the energy efficiency score of the DWTP evaluated, M is the number of inputs considered, S is the number of outputs produced, N is the number of DWTPs evaluated, and λ_k is a set of intensity variables representing the weighting of each DWTP evaluated, k, in the composition of the efficient frontier.

The energy efficiency score (θ_j) ranges from 0 to 1. A DWTP is energy efficient when $\theta_j = 1$ and inefficient when $\theta_j < 1$. For an energy-inefficient DWTP, the value of $1 - \theta_j$ informs us about the potential for energy savings, as it is the proportional input (energy consumption in this study) that can be achieved by DWTP j, given the level of outputs produced.

2.2. Double-Bootstrap DEA Model

The double-bootstrap DEA model proposed by Simar and Wilson [15] is based on the simulation of sample distribution by mimicking of the data-generation process (DGP). Assuming that the original data sample was generated by the DGP, energy efficiency scores are re-computed with the simulated data [19]. In other words, the bootstrapping procedure generates *new* data that is used to re-estimate energy efficiency scores using Equation (1). Then, the distinction between the *true* and *estimated* frontiers allows for statistical inference in DEA, i.e., for the identification of determinants of energy efficiency [20].

The double-bootstrap procedure employed in this study is Algorithm 2 of Simar and Wilson's model [20], summarized as follows:

1. Estimate the energy efficiency scores, θ_j, for all DWTPs in the sample using Equation (1).
2. Carry out a truncated maximum likelihood estimation to regress energy efficiency scores against a set of explanatory variables b_j, $\theta_j = b_j\beta + \varepsilon_j$, and provide an estimate $\hat{\beta}$ of the coefficient vector β and estimate $\hat{\sigma_\varepsilon}$ of σ_ε, the standard deviation of the residual errors ε_j.
3. For each DWTP, j ($j = 1, \ldots, N$), repeat the following four steps (3.1–3.4) Z_1 times to obtain a set of Z_1 bootstrap estimates $\hat{\theta_{jz}}$ for $z = 1, \ldots, Z_1$.

    3.1 Generate the residual error, ε_j, from the normal distribution $N(0, \hat{\sigma_\varepsilon^2})$.

    3.2 Compute $\theta_j^* = b_j\hat{\beta} + \varepsilon_j$.

    3.3 Generate a pseudo data set (x_j^*, y_j^*) where $y_j^* = y_j(\frac{\theta_j}{\theta_j^*})$.

    3.4 Using the pseudo data set (x_j^*, y_j^*) and Equation (1), calculate the pseudo energy efficiency estimates, $\hat{\theta_j^*}$.

4. Calculate the bias-corrected estimator, $\hat{\theta}_j$, for each DWTP, j ($j = 1, \ldots, N$), using the bootstrap estimator or the bias $\hat{z}_j$ where $\hat{\theta}_j = \theta_j - \hat{z}_j$ and $\hat{z}_j = \left(\frac{1}{Z_1}\sum_{z=1}^{Z_1}\hat{\theta_{jz}^*}\right) - \theta_j$.
5. Use the truncated maximum likelihood estimation to regress $\hat{\theta}_j$ on the explanatory variables, b_j, and provide an estimate $\hat{\beta^*}$ for β and an estimate $\hat{\sigma^*}$ for σ_ε.
6. Repeat the following three steps (6.1–6.3) Z_2 times to obtain a set of Z_2 pairs of bootstrap estimates $(\hat{\beta_j^{**}}, \hat{\sigma_j^{**}})$ for $z = 1, \ldots, Z_2$.

    6.1 Generate the residual error ε_j from the normal distribution $N(0, \hat{\sigma^{*2}})$.

    6.2 Calculate $\hat{\theta_j^{**}} = b_j\hat{\beta^*} + \varepsilon_j$.

    6.3 Use the truncated maximum likelihood estimation to regress $\hat{\theta_j^{**}}$ on the explanatory variables, b_j, and provide an estimate $\hat{\beta^{**}}$ for β and an estimate $\hat{\sigma^{**}}$ for σ_ε.

7. Construct the estimated $(1-\alpha)\%$ confidence interval of the n-th element, β_n, of the vector β, that is, $[Lower_{\alpha n}, upper_{\alpha n}] = [\hat{\beta_n^*} + \hat{a_\alpha}, \hat{\beta_n^*} + \hat{b_\alpha}]$ with $Prob\left(-\hat{b_\alpha} \le \hat{\beta_n^{**}} - \hat{\beta_n^*} \le -\hat{a_\alpha}\right) \approx 1 - \alpha$.

3. Sample and Variables

Chile is a long country whose extension is 4300 km. Hence, from a hydrometeorological point of view, Chile is a diverse country with significant precipitation variability throughout the country, increasing more than 500% from the North to the Austral region. Water runoff is also heterogeneous and varies from 510 m^3/person/year to 2,300,000 m^3/person/year. This diversity is expected to be accentuated by climate change, which will affect Chile in a complex fashion, with increased temperatures throughout the country and decreased precipitation in the Central and Southern areas of the country [21].

For empirical application, a sample of 146 Chilean DWTPs was used. All of the facilities were operated by private water companies, as the Chilean water industry was privatized between 1998 and 2004. The drinking water produced by all DWTPs must meet the quality standards of national norm NCh 409, which is based on the guidelines for drinking water quality published periodically by the World Health Organization (WHO) [22]. The quality of the drinking water supplied to citizens is monitored by water companies and the water regulation agency (Superintendencia de Servicios Sanitarios).

The energy consumed to treat raw water, expressed in kWh/year, was selected as the input; i.e., as the variable that DWTPs should minimize to improve energy performance. Following Molinos-Senante and Guzmán [10], the assessment of energy efficiency focused on DWTPs, without consideration of the energy used for groundwater pumping and raw water transport from the natural and artificial reservoirs to the DWTP.

In the assessment of water utility efficiency, the outputs selected should summarize the utilities' main function [23]. In this case study, the main function of DWTPs was considered to be the production of drinking water that met the NCh 409 quality standards. Moreover, the energy consumed by DWPTs (in kWh/year) depends not only on the volume of drinking water produced, but also on the pollutants removed, i.e., on the quality of the raw water and drinking water produced [12]. Hence, following the approach applied in previous studies [24,25] four quality-adjusted outputs (QAOs) were considered to assess DWTP energy efficiency, defined as follows:

$$QAO_p = V \cdot \frac{(C_{pin} - C_{pef})}{C_{pin}}, \tag{2}$$

where V is the volume of drinking water produced (in m^3/year), C_{pin} is the concentration of pollutant C_p in the influent of the DWTP, and C_{pef} is the concentration of pollutant C_p in the effluent of the DWTP. Molinos-Senante and Sala-Garrido [7] defined energy intensity as the "energy consumed (kWh) per unit volume (m^3) of drinking water produced". These authors developed energy intensity functions for DWTPs using parametric regression analysis to identify the main drivers of energy use in DWTPs. They concluded that energy intensity of DWTPs depended on the capacity of the facility and the removal efficiency of i) turbidity, ii) arsenic, iii) total dissolved solids, and iv) sulfates. Hence, these four pollutants were considered in this study as QAOs.

According to Sanders and Webber [26], the energy efficiency of a given water treatment technology correlates with the size, concentrations, and nature of the pollutants to be removed. Other factors, such as the water company operating the facility or the age of the DWTP, might also impact energy efficiency [9]. Based on previous studies [4,27], four variables were included as potential determinants of DWTP energy efficiency scores: i) DWTP age, ii) raw water source, iii) ownership of the company operating the DWTP, and iv) type of treatment.

Table 1 shows the main statistics for the variables (input, outputs, and factors underlying efficiency) used in this empirical application.

Table 1. Sample description. (Source: Water and Sewerage Management Reports by SISS).

Data	Variables	Average	SD	Minimum	Maximum
Input	Energy consumed (kWh/year)	258,878	441,353	1,152	2,331,739
	Energy consumed (kWh/m^3)	0.175	0.259	0.022	1.194
Outputs	Water treated (m^3/year)	17,306,317	61,159,647	6,642	386,483,247
	Efficiency in turbidity removal (%)	56.2	21.2	5.6	95.5
	Efficiency in arsenic removal (%)	58.2	24.4	4.8	86.7
	Efficiency in TDS removal (%)	39.2	24.3	4.8	80.0
	Efficiency in sulfates removal (%)	47.3	28.3	4.4	90.9
Continous explanatory variable	Age (years)	27.4	15.5	11.0	52.0
				Number	% total
Categorial explanatory variable	Source of water		Surface	45	30.8
			Groundwater	43	29.5
			Mixed	58	39.7
	Water company		Private operator	76	52.0
			Concession	70	48.0
	Type of treatment		PF	70	48.0
			RGF	76	52.0

SD: Standard Deviation; PF: Pressure filtration; RGF: Rapid gravity filtration.

4. Results and Discussion

4.1. Estimated Energy Efficiency

The energy efficiency of each DWTP was estimated following the methodological approach described in Section 2.1., i.e., by employing Equation (1). According to the energy efficiency scores calculated with the original data, 6 of 146 (4.1%) DWTPs were energy efficient (Figure 1). These DWTPs formed the best practice frontier, as they used the minimum quantity of energy given their efficiency of pollutant removal compared with the other DWTPs evaluated. The mean energy efficiency of the DWTPs assessed was 0.38, meaning that on average they could reduce the energy consumed by 62% while retaining output production if they were operated as energy-efficient facilities. The average energy efficiency seems to be low, but this result is consistent with those of previous studies. Molinos-Senante and Guzmán [10] and Hernández-Sancho et al. [28] reported average energy efficiencies of 0.45 and 0.31 for samples of Chilean DWTPs and Spanish WWTPs, respectively. The results of this study confirmed in general that the water treatment facilities had notable room for energy saving. Moreover, the sample of DWTPs evaluated was very heterogeneous in terms of energy efficiency (Figure 1); almost one-third (42 of 146) of the facilities had energy efficiency scores <0.2, indicating dramatically poor energy performance and thus much improvement potential.

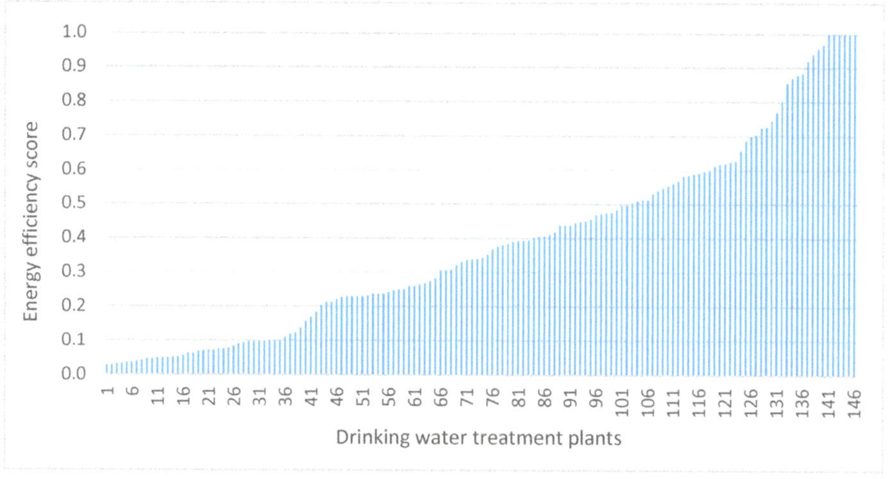

Figure 1. Original energy efficiency scores of drinking water treatment plants.

As reported in Section 2.2, to compute bias-corrected energy efficiency scores and bootstrap 95% confidence intervals, 2000 bootstrap samples were generated. The original and bias-corrected bootstrap energy efficiency estimates are compared in Figure 2. To facilitate comparison, only results for the 30 DWTPs with the highest original energy efficiency scores are presented. Detailed information for all DWTPs evaluated is provided in the Supplementary Materials. As expected from a theoretical point of view [12,16,29], the bias signs were negative (bias-corrected energy efficiency scores were lower than original scores) for the 146 DWTPs assessed. The average bias-corrected energy efficiency score was 0.28, meaning that DWTPs could conserve 71% of the energy currently used while maintaining output generation if they were operated as energy-efficient facilities. The difference between the bias-corrected and original energy efficiency scores reflects the limitations of the traditional DEA model, which does not integrate data variability in efficiency assessment. Although the difference in average scores was not large, it altered the ranking of DWTPs (Figure 2). For example, DWTP 144 was ranked first based on the original energy efficiency score (1), but 15th of the 30 DWTPs listed in Figure 2 based on the bias-corrected score. Under both methodological approaches, DWTP 141 was ranked first, showing that it was the most energy efficient facility in the sample. It is one of the largest facilities evaluated and it employs rapid gravity filters to treat raw water. To verify that the DWTP ranking differed statistically according to the DEA model used, the non-parametric Mann–Whitney test was performed. The p value was <0.01, reflecting a strongly significant difference in ranking.

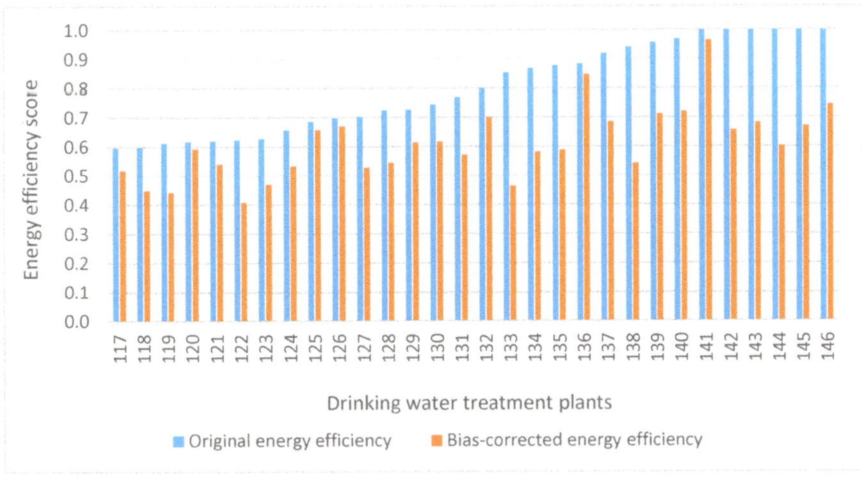

Figure 2. Ranking of drinking water treatment plants based on original and bias-corrected energy efficiency scores.

The lower and upper bounds are the maximum and minimum energy efficiency scores computed, considering the 2000 bootstrap samples generated. In other words, the difference between the lower and upper bounds represented the variability in energy efficiency for each DWTP evaluated. For DWTPs with the lowest energy efficiency scores, the gaps between the upper and lower bounds were small (minimum, 0.008), reflecting almost no variability (Figure 3). By contrast, the gaps between maximum and minimum energy efficiency scores was large (maximum, 0.75) for DWTPs with the highest original energy efficiency scores. This finding showed the importance of considering data variability in efficiency assessment employing the DEA approach, to provide more reliable and robust energy efficiency estimations to support decision-making. Such integration of variability is essential when the purpose of the analysis is to rank facilities based on performance (Figure 3). In the context of regulated water industries, this issue is very relevant, as benchmarking is used to set water tariffs in several water regulation models.

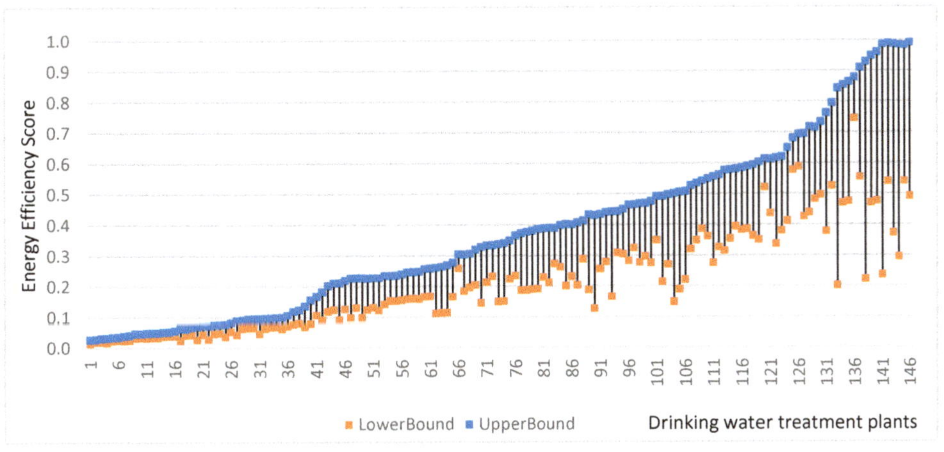

Figure 3. Lower and upper bounds of energy efficiency for each drinking water treatment plant.

4.2. Determinants of Energy Efficiency

To improve the energy efficiency of DWTPs, not only inefficient facilities, but also the factors that influence energy efficiency, must be identified. This is the main advantage of the double-bootstrap technique. In the second stage of analysis, regression was conducted with four variables to identify factors influencing DWTP energy efficiency. Table 2 shows the bias-corrected coefficients of the regressed variables, with standard errors and p values. From a statistical point of view, a variable influences the energy efficiency of DWTPs at the 95% significance level if its p value is ≤ 0.05.

Table 2. Results of bootstrap regression.

Variables	Bias-Corrected Coefficient	Standard Error	p-Value
Age	−25.25	4.51	0.030 **
Volume	10.44	2.14	<0.001 *
Surface water	Reference variable		
Groundwater	58.54	10.25	0.751
Mixed water	45.32	7.55	0.354
Privatized company	Reference variable		
Concessionary company	10.25	3.24	0.548
Pressure filtration	Reference variable		
Rapid gravity filtration	−5.25	1.21	0.006 *

* Significant at 1% level; ** Significant at 5% level.

The DWTP age positively influenced energy efficiency (Table 2). Hence, older water treatment facilities exhibited a better energy performance than did younger facilities. The oldest facility analyzed had been operating for 52 years, and the sample contained a non-minor number of DWTPs that were more than 25 years old, in which old equipment had been replaced with newer, more efficient systems. This finding reflects the importance of proper equipment maintenance and the continuous incorporation of processes to improve the energy efficiency of DWTPs [10].

As in the case of WWTPs [15], the water treatment facilities presented economies of scale regarding energy use (Table 2). Larger DWTPs had significantly higher energy efficiency scores than did smaller facilities. This information is essential for the planning of new DWTPs, given the current tendency to decentralize urban water treatment facilities to increase redundancy in the case of an unplanned event (e.g., earthquake, volcanic eruption, or hurricane). However, from economic and environmental perspectives, larger DWTPs, i.e., centralized systems, are more favorable because per-unit energy use decreases with increasing capacity.

The quality of raw water, and thus the treatment intensity required to produce drinking water, sometimes depends on the water source. The DWTPs evaluated treated surface water, groundwater, and mixed water (Table 1). Following [12], the raw water source variable was integrated into the regression analysis as three dummy variables. The results showed that the source of raw water did not affect DWTP energy efficiency. This finding was consistent with the definition of the outputs considered in energy efficiency estimation, which included the concentrations of pollutants in DWTP influents and effluents. Usually, the use of groundwater requires more energy for water pumping, as well as the depth, which depends on water availability. However, in this case study, to guarantee homogeneity to the greatest extent possible, energy use for groundwater pumping was not considered in the analysis.

Several studies [24,30,31] have focused on the comparison of the performance of public and private water companies. In Chile, 98% of urban customers are supplied by private water companies [22], which operate under two regimes: i) Fully private water companies following the English and Welsh model, and ii) concessionary water companies following the French model [32] (The difference between water company types is the concession term (perpetuity for fully private companies and 30 years for concessionary water companies). Thus, DWTP ownership was integrated in the analysis as two artificial dummy variables. DWTP ownership did not impact energy efficiency (Table 2). This finding was consistent with Sowby's [33] finding that energy efficiency did not differ significantly between public and private water companies.

For WWTPs, which have been studied more widely than DWTPs, results regarding the influence of treatment technologies on energy efficiency were inconclusive [34]. Thus, we investigated whether the technology used (i.e., pressure filtration or rapid gravity filtration) was a determinant factor for DWTP energy efficiency. The average energy efficiency score for water treatment facilities using pressure filtration was 0.351, and that for those employing rapid gravity filtration was 0.408; thus, the latter facilities exhibited significantly better energy performance (Table 2).

From the perspective of cleaner production, the findings of this study demonstrate the importance of adequate maintenance and equipment replacement to ensure DWTP energy efficiency. Moreover, the role of the technology in this efficiency has been revealed. This issue is fundamental for decision-making, especially in the context of the UN's Sustainable Development Goals. The achievement of Goal 6 ("by 2030, achieve universal and equitable access to safe and affordable drinking water for all") will involve the construction and operation of new DWTPs. The technological factor must be taken into account to reduce the energy requirements of these new facilities.

5. Conclusions

The assessment of energy requirements for urban water supplies has emerged as a relevant topic. The few previous studies that have evaluated the energy efficiency of DWTPs have employed conventional DEA models, ignoring the deterministic nature of this method. To overcome this limitation, a double-bootstrap DEA model was used in this study to evaluate the energy efficiency of a sample of DWTPs.

Empirical application with a sample of Chilean DWTPs provided three primary conclusions. First, the energy efficiency of the sample of DWTPs evaluated was very low; less than 5% of facilities were energy efficient, and the DWTPs could reduce the energy consumed by >50% while maintaining the same level of pollutant removal. Many water companies have focused on the optimization of energy use in WWTPs in recent years; the findings of this study show that improving the energy efficiency of DWTPs is also challenging. Second, the integration of data variability in the energy efficiency assessment notably affected the results. The ranking of DWTPs based on original and bias-corrected energy efficiency scores differed significantly. Water regulators that use benchmarking to regulate water companies or to set water tariffs should integrate data variability in their performance assessments, as in this study, to avoid the generation of biased results and conclusions. Third, among the variables studied, the determinants of energy efficiency were the volume of raw water treated (i.e., facility capacity), DWTP age, and the main technology used to treat raw water (i.e., pressure filters or rapid gravity filters). Plant size and the technology used to treat raw water are structural variables that cannot be modified by DWTP managers; thus, the improvement of energy efficiency in existing facilities is difficult in the short term. However, in consideration of long-term energy efficiency, these features must be taken into account when planning the construction of new DWTPs. In addition, facility age positively influenced energy efficiency, revealing the important role of equipment maintenance and replacement in the energy efficiency of water treatment facilities.

The improvement of DWTP energy efficiency is essential to achieve global climate goals and provide affordable drinking water for all people. Several regulations, such as the Revised Energy Efficiency Directive (which defines the European Union energy efficiency targets), and international agreements are focused on the reduction of consumers' energy requirements. The European Water Framework Directive, which establishes full cost recovery for water services, was implemented in 2000. It mandates that all costs of urban water services, including DWTPs, must be transferred to citizens via water tariffs. The improvement of DWTP energy efficiency contributes to the achievement of the UN's Sustainable Development Goal 6 and to the fulfillment of climate change agreements. In this context, to support decision-making, it is essential not only to assess the energy efficiency of DWTPs, but also to identify the determinants of energy efficiency using a reliable and robust methodological approach, as was done in this study.

Supplementary Materials: The Supplementary Materials are available online at http://www.mdpi.com/1996-1073/12/4/765/s1.

Author Contributions: Conceptualization, M.M.-S. and R.S.-G.; methodology, R.S.-G.; software, R.S.-G.; validation, M.M.-S. and R.S.-G.; formal analysis, M.M.-S.; investigation, M.M.-S. and R.S.-G.; resources, M.M.-S.; data curation, R.S.-G.; writing—original draft preparation, M.M.-S.; writing—review and editing, M.M.-S.; visualization, R.S.-G.; supervision, M.M.-S. and R.S.-G.; project administration, M.M.-S.; funding acquisition, M.M.-S.

Funding: This research was funded by CONICYT, programme Fondecyt (11160031).

Conflicts of Interest: The authors declare no conflict of interest.

References

1. Lam, K.L.; Kenway, S.J.; Lant, P.A. Energy use for water provision in cities. *J. Clean. Prod.* **2017**, *143*, 699–709. [CrossRef]
2. Wang, S.; Cao, T.; Chen, B. Urban energy-water nexus based on modified input-output analysis. *Appl. Energy* **2016**, *196*, 208–217. [CrossRef]
3. Wakeel, M.; Chen, B.; Hayat, T.; Alsaedi, A.; Ahmad, B. Energy consumption for water use cycles in different countries: A review. *Appl. Energy* **2016**, *178*, 868–885. [CrossRef]
4. Miller, L.A.; Ramaswami, A.; Ranjan, R. Contribution of water and wastewater infrastructures to urban energy metabolism and greenhouse gas emissions in cities in India. *J. Environ. Eng.* **2013**, *139*, 738–745. [CrossRef]
5. Lee, M.; Keller, A.A.; Chiang, P.-C.; Den, W.; Wang, H.; Hou, C.-H.; Wu, J.; Wang, X.; Yan, J. Water-energy nexus for urban water systems: A comparative review on energy intensity and environmental impacts in relation to global water risks. *Appl. Energy* **2017**, *205*, 589–601. [CrossRef]
6. Santana, M.V.E.; Zhang, Q.; Mihelcic, J.R. Influence of water quality on the embodied energy of drinking water treatment. *Environ. Sci. Technol.* **2014**, *48*, 3084–3091. [CrossRef] [PubMed]
7. Molinos-Senante, M.; Sala-Garrido, R. Energy intensity of treating drinking water: Understanding the influence of factors. *Appl. Energy* **2017**, *202*, 275–281. [CrossRef]
8. Sowby, R.B.; Burian, S.J. Survey of energy requirements for public water supply in the United States. *J. Am. Water Works Assoc.* **2017**, *109*, E320–E330. [CrossRef]
9. Molinos-Senante, M.; Sala-Garrido, R. Evaluation of energy performance of drinking water treatment plants: Use of energy intensity and energy efficiency metrics. *Appl. Energy* **2018**, *229*, 1095–1102. [CrossRef]
10. Molinos-Senante, M.; Guzmán, C. Benchmarking energy efficiency in drinking water treatment plants: Quantification of potential savings. *J. Clean. Prod.* **2018**, *176*, 417–425. [CrossRef]
11. Ananda, J. Productivity implications of the water-energy-emissions nexus: An empirical analysis of the drinking water and wastewater sector. *J. Clean. Prod.* **2018**, *196*, 1097–1105. [CrossRef]
12. Ananda, J. Explaining the environmental efficiency of drinking water and wastewater utilities. *Sustain. Prod. Consum.* **2019**, *17*, 188–195. [CrossRef]
13. Marques, R.C.; Berg, S.; Yane, S. Nonparametric benchmarking of Japanese water utilities: Institutional and environmental factors affecting efficiency. *J. Water Resour. Plan. Manag.* **2014**, *140*, 562–571. [CrossRef]
14. See, K.F. Exploring and analysing sources of technical efficiency in water supply services: Some evidence from Southeast Asian public water utilities. *Water Resour. Econ.* **2015**, *9*, 23–44. [CrossRef]
15. Cazals, C.; Florens, J.-P.; Simar, L. Nonparametric frontier estimation: A robust approach. *J. Econ.* **2002**, *106*, 1–25. [CrossRef]
16. De Witte, K.; Marques, R.C. Influential observations in frontier models, a robust non-oriented approach to the water sector. *Ann. Oper. Res.* **2010**, *181*, 377–392. [CrossRef]
17. SISS. Drinking Water and Sewerage Management Report in Chile. 2016. Available online: http://www.siss.gob.cl/586/w3-propertyvalue-6415.html (accessed on 10 October 2018).
18. Charnes, A.; Cooper, W.W.; Rhodes, E.L. Measuring the Efficiency of Decision Making Units. *Eur. J. Oper. Res.* **1978**, *2*, 429–444. [CrossRef]

19. Banker, R.D.; Charnes, A.; Cooper, W.W. Some models for estimating technical and scale inefficiencies in data envelopment analysis. *Manag. Sci.* **1984**, *30*, 1078–1092. [CrossRef]
20. Simar, L.; Wilson, P.W. Estimation and inference in two-stage, semi-parametric models of production processes. *J. Econ.* **2007**, *136*, 31–64. [CrossRef]
21. Güngör-Demirci, G.; Lee, J.; Keck, J. Assessing the performance of a California water utility using two-stage data envelopment analysis. *J. Water Resour. Plan. Manag.* **2018**, *144*, 05018004. [CrossRef]
22. Cai, Y.; Sam, C.Y.; Chang, T. Nexus between clean energy consumption, economic growth and CO_2 emissions. *J. Clean. Prod.* **2018**, *182*, 1001–1011. [CrossRef]
23. Ananda, J. Evaluating the performance of urban water utilities: Robust nonparametric approach. *J. Water Resour. Plan. Manag.* **2014**, *140*, 04014021. [CrossRef]
24. Donoso, G. *Water Policy in Chile*; Global Issues in Water Policy; Springer: Cham, Switzerland, 2018.
25. Lorenzo-Toja, Y.; Vázquez-Rowe, I.; Chenel, S.; Moreira, M.T.; Feijoo, G. Eco-efficiency analysis of Spanish WWTPs using the LCA+DEA method. *Water Res.* **2015**, *68*, 651–666. [CrossRef] [PubMed]
26. Saal, D.S.; Parker, D.; Weyman-Jones, T. Determining the contribution of technical change, efficiency change and scale change to productivity growth in the privatized English and Welsh water and sewerage industry: 1985–2000. *J. Product. Anal.* **2007**, *28*, 127–139. [CrossRef]
27. Dong, X.; Zhang, X.; Zeng, S. Measuring and explaining eco-efficiencies of wastewater treatment plants in China: An uncertainty analysis perspective. *Water Res.* **2017**, *112*, 195–207. [CrossRef] [PubMed]
28. Sanders, K.T.; Webber, M.E. Evaluating the energy consumed for water use in the United States. *Environ. Res. Lett.* **2012**, *7*, 034034. [CrossRef]
29. Gude, V.G. Energy and water autarky of wastewater treatment and power generation systems. *Renew. Sustain. Energy Rev.* **2015**, *45*, 52–68. [CrossRef]
30. Hernández-Sancho, F.; Molinos-Senante, M.; Sala-Garrido, R. Energy efficiency in Spanish wastewater treatment plants: A non-radial DEA approach. *Sci. Total Environ.* **2011**, *409*, 2693–2699. [CrossRef] [PubMed]
31. Simões, P.; Marques, R.C. Performance and congestion analysis of the portuguese hospital services. *Cent. Eur. J. Oper. Res.* **2011**, *19*, 39–63. [CrossRef]
32. Suárez-Varela, M.; de los Ángeles García-Valiñas, M.; González-Gómez, F.; Picazo-Tadeo, A.J. Ownership and Performance in Water Services Revisited: Does Private Management Really Outperform Public? *Water Resour. Manag.* **2017**, *31*, 2355–2373. [CrossRef]
33. Guerrini, A.; Molinos-Senante, M.; Romano, G. Italian regulatory reform and water utility performance: An impact analysis. *Util. Policy* **2018**, *52*, 95–102. [CrossRef]
34. Marques, R.C. *Regulation of Water and Wastewater Services*; An International Comparison; IWA Publishing: London, UK, 2010.

Article

Estimating Relative Efficiency of Electricity Consumption in 42 Countries during the Period of 2008–2017

Chia-Nan Wang [1,2], Quoc-Chien Luu [1,*] and Thi-Kim-Lien Nguyen [1]

1 Department of Industrial Engineering and Management, National Kaohsiung University of Science and Technology, Kaohsiung 80778, Taiwan; cn.wang@nkust.edu.tw (C.-N.W.); lien.nguyen0209@gmail.com (T.-K.-L.N.)

2 Department of Industrial Engineering and Management, Fortune Institute of Technology, Kaohsiung 83160, Taiwan

* Correspondence: jkie2211@gmail.com; Tel.: +886-097-965-2251

Received: 26 September 2018; Accepted: 31 October 2018; Published: 5 November 2018

Abstract: Augmentation of electrical equipment is pushing for an increase in energy supply sources all over the world, as electricity consumption (EC) typically rises with growing populations. The value of EC reveals economic development and degree of emissions. Therefore, this research uses the undesirable outputs model in data envelopment analysis (DEA) for estimating relative efficiency of electricity consumption in 42 countries from 2008 to 2017. According to the principle of an undesirable outputs model and studied objectives, variables are selected that included population and EC as inputs, gross domestic product (GDP) as desirable output, and carbon dioxide (CO_2), methane (CH_4), and nitrous oxide (N_2O) as undesirable outputs. The empirical results indicate that 420 terms of 42 countries during the period of 2008–2017 have 102 efficient and 310 inefficient terms. Moreover, the interplay level between input and output factors every year is presented via scores. The study suggests the effect of EC to human life and propounds the emission status to look for directions to overcome inefficient terms.

Keywords: electricity consumption (EC); undesirable outputs model; data envelopment analysis (DEA); efficient; inefficient

1. Introduction

In modern life, electrical energy is essential to meet the demands of extending technology and electronic equipment [1], as electricity provides energy for lighting, heating, cooling, factories, machines, transportation systems, i.e., [2]. The increasing population leads to increasing electricity consumption (EC); thus, population growth and EC have a significant positive relationship [3]. When electricity is utilized, it contributes to enhancing the economic development index. Lu indicated that a 1% increase in EC from 17 Taiwanese industries boosted the real GDP by 1.72% [4]; Enu and Patrick explained the effect of EC on economic growth in Ghana [5]; Altisnay and Karagol showed the casual relationship between EC and real GDP in Turkey [6]. On the other hand, EC causes pollutant emissions to the environment, including CO_2, CH_4, and N_2O. For instance, a study by the Federal University of Agriculture Abeokuta assessing carbon footprints over the 2011–2012 period showed that 5935 tons of CO_2 represented 63% transportation, 35% campus energy consumption, and 2% farm machineries per student [7]. In Hong Kong, between 2002 and 2015, the annual EC went from 27 to 34.1 million tons; further, CO_2-eq/kWh was increased from 702 to 792 g [8]. Therefore, EC has a positive and significant relationship with both emissions [9,10] and economic growth [11].

Electricity is generated from two sources, i.e., nonrenewable and renewable energy. Renewable energy comprises hydropower, biomass, wind, solar, and geothermal. Nonrenewable energy consists of oil, natural gas, coal, and nuclear. Both sources are applied to generating electricity to provide energy for inhabitants and their applications [12]. The population increase augments the EC as well. When a consumer uses electrical energy, economic growth is extended, and CO_2, CH_4, and N_2O rise as well. Increased emissions lead to polluted environments and climate change. Thus, the purpose of study is to determine the relationship among inputs (population, and EC), desirable output (GDP), and undesirable outputs (emissions), the relation is evaluated via the scores computed by an undesirable outputs model in DEA.

In DEA, the super-SBM, EBM, and Malmquist models can formulate the maximum score and separate values for each decision-making unit (DMU) in every term; however, they cannot deal with desirable and undesirable outputs, whereas an undesirable outputs model only approaches to the highest value of 1, but it can solve with good (desirable) and bad (undesirable) outputs independently [13]. This model reaches bad factors in the operation process; the inefficient DMUs will be suggested, i.e., raising good outputs while simultaneously reducing bad outputs to improve their scores [14,15]. With these characteristics, the study applies an undesirable outputs model into computing the efficiency of EC with its relative elements in 42 countries over the world from 2008 to 2017. The analysis result works out the influence of EC on the economic development, and emissions in which the increased levels of undesirable emissions are the root causes of climate change. A feasible solution is recommended to refine the performance of inefficient terms. Moreover, the study draws a picture of the productivity efficiency between EC and its relative factors in 42 countries over the years.

The study is arranged as follows: Section 1 shows the general points of electricity's application, producing an electricity process, and its effects; Section 2 overviews EC and its background research, the theoretical concept of undesirable outputs model in undesirable model and its application; Section 3 builds upon the proposal research and methodology, and quotes source materials; Section 4 displays the empirical analysis results; Section 5 comments on the general results, gives limitations, and discusses future research.

2. Literature Review

The life of people without electricity was inconvenient, they worked by manual labour, and lived without light. Since electrical energy was invented in the18th century, the life of inhabitants has been changed with access to light, electronic equipment, and high-tech. The effectiveness of production operations is enhanced and upgraded sharply by the use of electrical machines. The process whereby people use electrical energy for lighting, heating, transportation, and so on is called "electricity consumption". The population is the major source that supports electricity development when utilizing electronic equipment. The electricity is consumed at a high or low level, the EC reflects an economic growth level. Chen denoted that the economic growth and population have a vital role on the electrical energy consumption when depending on the non-parametric model [16]. To explore the electricity demand in the future, Gajowniczek [17] displayed an approach to predict electricity load at the individual household level using CART, SVM and a MLP neutral network model; Gajowniczek continued studying electricity demand [18]; Singh [19] proposed Bayesian network prediction for energy usage forecasting.

On the other hand, the electricity causes greenhouse gas [20] that leads to climate change because of the emission of CO_2, CH_4, N_2O [21] from electricity production processes [22]. Emissions from hydropower are estimated by using statistical global emission models through the reservoir water surface [23,24], that from natural gas and coal power plants is calculated by a simple model [25], and that from combustion power plants is counted by the values and data of emission factors exhausted from the circulating fluidized bed boiler [26], or that from wind power plants is formulated by a simple analysis method for the undesirable elements of electricity production processes [27]. In China emissions from EC are determined by a data analysis and measurement method [28], while in the United States a transparent method is used [29]. Hence, the previous researchers applied various methods to an examination of the emission of undesirable factors from EC.

Whereas DEA normally concerns calculating performance with the inputs and good outputs in various models such as dynamic-SBM, super-SBM, EBM, i.e., however, they cannot solve for undesirable outputs in social activities, air pollution, and the industrial manufacturing sector. For this reason, Tone proposed an undesirable outputs model in DEA to evaluate bad outputs [30], displaying a new scheme. A DMU acquires efficiency as the score approaches 1, and it is inefficient when the score is less than 1. Furthermore, the model can compute the performance by combining both undesirable and desirable outputs [31]. The efficiency valuations indicate not only the interplay between desirable and undesirable outputs, but also the ranking of each DMU in every year [32]. Many researchers have applied the undesirable model into their studies. For example, an analysis by the Organization for Economic Co-operation and Development (OECD) of countries with population and energy consumption as input factors, GDP as desirable output, and CO_2 as undesirable output reveals the environmental efficiency [33]; the overall efficiency of the United States's electricity production is evaluated by escalating the desirable output and undesirable outputs [34]; counting the efficiency shows the relationship between labor force, energy consumption, government expenditure as input, GDP as desirable output, and CO_2 emissions as undesirable output [35]. Moreover, the undesirable model is used for examining performance in other aspects such as estimating the impact of production pollutants in the textile industry of China with inputs like labor, and energy, yam and fabric as desirable outputa, and wastewater as undesirable [36]. In addition, the researchers also utilized an undesirable model to analyze and evaluate efficiency in the energy sector. Measuring between inputs including gross fixed capital formation, labor and energy consumption and outputs including CO_2 (undesirable output), and GDP (desirable output) indicated the energy performance in Brazil, Russia, India, China, and South Africa [37].

With the principle of the undesirable outputs model and its previous applications, the paper proposed undesirable outputs model of DEA to analyze the interplay between inputs such as population and EC and outputs such as GDP, CO_2, CH_4, N_2O in the electricity production aspect of 42 countries during the 2008–2017 term.

3. Methodology

3.1. Proposal Research

Our study of the electricity performance process in 42 countries is organized into four steps as shown in Figure 1:

- Step 1: Present the purpose of the selected topic, input, and output variables. The theme and data must be reselected if they are inappropriate. The suitable materials of electricity, as listed on Enerdata [38], Worldbank [39], and Epa [40], are collected. Then, the EC from all over the world is introduced and factors relating the production process with EC are described.
- Step 2: Show the benefits of electricity. The study overviews EC and its influences on the environment; and the undesirable model theory is used to demonstrate in feasibility of the method. Especially, previous studies that relate to EC and the undesirable model to indicate a probability theme are discussed.
- Step 3: The first stage of the analysis process must check the Pearson coefficient to ensure the data is isotonic; any value does not range from −1 to +1 it must be removed and reselected. Next, the suitable values are applied into an undesirable outputs model to compute scores. The scores are used for determining the efficiency/inefficiency of 42 countries over the years. The scores propound their ranking over each term as well. The empirical results present a stable or upward and downward interplay of countries during the period of 2008–2017 in particular. Moreover, the analysis results suggest the current status of the effect level in each year when utilizing electrical energy.
- Step 4: Manifest main points of the empirical results of efficient/inefficient countries, and ranking, in addition to recommendations on the analysis of a variable pathway of each country in every year. The suggestion points out improvements for inefficient countries.

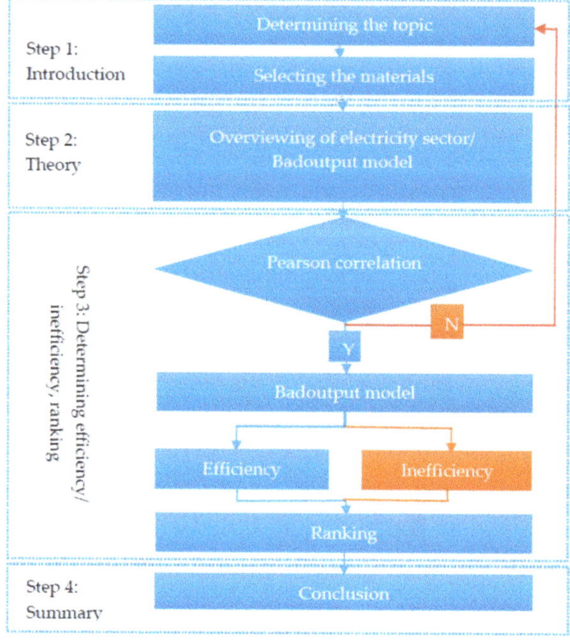

Figure 1. Proposal research.

3.2. Data Source

Electricity is a source fuel that provides lighting, heating, cooling, and runs electronics, machinery, and transportation systems. Hence, in modern life with the increasing use of diverse high-tech and electrical equipment, electricity is an essential element. While on the subject, the research discovers electricity consumption levels and their relative factors. Based on the input and output data posted on websites, including electricity consumption on Enerdata [38], population and GDP on Worldbank [39], emissions, including CO_2, CH_4, and N_2O, are computed when their equations are based on the Epa version 3.2 of June 2014 [40]. The 42 countries selected from Enerdata [38] to estimate the performance as listed in Table 1.

Table 1. Name of countries.

No	Country	No	Country	No	Country
1	Belgium	15	Kazakhstan	29	Japan
2	Czech Republic	16	Russia	30	Malaysia
3	France	17	Ukraine	31	South Korea
4	Germany	18	Uzbekistan	32	Thailand
5	Italy	19	Canada	33	Australia
6	Netherlands	20	United States	34	New Zealand
7	Poland	21	Argentina	35	Algeria
8	Portugal	22	Brazil	36	Egypt
9	Romania	23	Chile	37	Nigeria
10	Spain	24	Colombia	38	South Africa
11	Sweden	25	Mexico	39	Iran
12	United Kingdom	26	China	40	Kuwait
13	Norway	27	India	41	Saudi Arabia
14	Turkey	28	Indonesia	42	United Arab Emirates

Source: Enerdata [38].

Characteristics of each variable are described as follows:

- Population (Input): When the population of a nation increases, the electricity usage increases because the amount of electronic equipment will be augmented as well.
- Electricity consumption (Input): The electricity is consumed by providing electrical energy for light, heating, cooling, machines, and so on.
- GDP (desirable output): The economic performance of every country is measured by market value. In the electricity sector, the volumes of EC are used by consumers for any application, i.e., they contribute to extending GDP indicators.
- CO_2, CH_4, N_2O (undesirable outputs): Coal, oil, natural gas, and biomass are burned in combustion power plants. Nuclear power plants create heat, in addition to the heat of the Sun in solar power, turbines in hydropower plants via the energy power of water from natural waterfalls, tides, and flowing rivers create electricity, or turbines in wind power plants by the wind's energy. These processes all generate electricity, then the generation electricity is transmitted to customers via wires. When the electrical energy is consumed, the EC process produces emissions, including CO_2, CH_4, and N_2O.

3.3. Undesirable Outputs Model

The undesirable outputs model is utilized to calculate the performance of DMUs when its outputs obtain undesirable outputs. In this study, the undesirable outputs model is applied to deal with good (desirable) and (bad) (undesirable) outputs. We utilize an undesirable outputs model to compute the efficiency of the electrical energy consumption in 42 countries. The DMUs are the 42 countries, these countries are set up n DMU (a_0, b_0) ($n = 1, 2, \ldots, s$). Let the input factor be A, desirable factor (B^d), and undesirable factor (B^u). Then, the production possibility is given by:

$$P = \left\{ \left(a, b^d, b^u\right), a \geq X\lambda; b^d \leq B^d\lambda; b^u \geq B^u\lambda; L \leq e\lambda \leq U, \lambda \geq 0 \right\} \tag{1}$$

The intensity vector is λ, it means that the above definition corresponds to the constant return to scale technology [41], and the lower and upper bounds of the intensity vector are L and U, respectively ($e = (1, \ldots 1) \in R^+, L \leq 1, U \geq 1$). There is at least one strict inequality when formulating the efficiency of one DMU (a_0, b_0^d, b_0^u) without vector $(a_0, b_0^d, b_0^u) \in P$ and $a_0 \geq a, b_0^d \leq b^d, b_0^u \geq b^u$. According to the SBM of Tone [42], the objective function of the undesirable model is formulated as follows:

$$\rho^* = \min \frac{1 - \frac{1}{k}\sum_{i=1}^{k} \frac{s_i^-}{a_{i0}}}{1 + \frac{1}{s}\left(\sum_{r=1}^{s_1} \frac{s_r^d}{b_{ro}^d} + \sum_{r=2}^{s_1} \frac{s_r^u}{b_{ro}^u}\right)} \tag{2}$$

Subject to:

$$\begin{aligned} a_0 &= A\lambda + s^- \\ b_0^d &= B\lambda - s^d \\ b_0^u &= B\lambda + s^u \\ s^-, s^d, s^u, \lambda &\geq 0. \end{aligned}$$

The excess in inputs, bad outputs and shortages in good outputs are s^-, s^u, s^d, respectively. The number of factors in s^u and s^d are s_1 and s_2, respectively, and $s = s_1 + s_2$. Using an optimal solution as ρ^*, s^{-*}, s^{d*} and s^{u*} for determining the efficiency of country by undesirable outputs when $\rho^* = 1$, $s^{-*} = 0$, $s^{d*} = 0$, and $s^{u*} = 0$. When the DMU is inefficient, ρ^* can be improved in order to become efficient by moving the excesses in inputs and bad outputs, simultaneously increasing the shortfalls in good outputs [42] as follows:

$$\begin{aligned} a_0 - s^{-*} &\Rightarrow a_0 \\ b_0^d + s^{d*} &\Rightarrow b_0^d \\ b_0^u - s^{u*} &\Rightarrow b_0^u \end{aligned} \tag{3}$$

The above program was transformed into an equivalent linear program by Charnes and Cooper [43]. Let the dual variable vectors be x, y^d, y^u. Based on the dual side of the linear program, the dual program in the variable x, y^d, y^u for constant return to scale [30] is defined as below:

$$\max y^d b_0^d - x a_0 - y^u b_0^u. \tag{4}$$

Subject to:

$$\begin{aligned} & y^d B^d - xA - y^u B^u \leq 0 \\ & x \geq \frac{1}{k}\left[\frac{1}{a_0}\right] \\ & y^d \geq \frac{1 + y^d b_0^d - x a_0 - y^u b_0^u}{s}\left[\frac{1}{b_0^d}\right] \\ & y^u \geq \frac{1 + y^d b_0^d - x a_0 - y^u b_0^u}{s}\left[\frac{1}{b_0^u}\right] \end{aligned}$$

The virtual prices of inputs, desirable and undesirable outputs are replaced by the dual variables x, y^d, y^u respectively. The profit $y^d b^d - xa - y^u b^u$ [30] does not exceed zero for every DMU, and the profit $y^d b_0^d - x a_0 - y^u b_0^u$ for the DMU concerned when the dual program aims at obtaining the optimal virtual costs and prices for each DMU.

In addition, we set $w_1 \in R^+, w_2 \in R^+$ as the weights of desirable and undesirable outputs, respectively. The weights of bad and good outputs are converted to relative weights with their mathematical expression [30] as follows:

$$\rho^* = \min \frac{1 - \frac{1}{k}\sum_{i=1}^{k} \frac{s_{io}^-}{a_{io}}}{1 + \frac{1}{k}\left(W_1 \sum_{r=1}^{s_1} \frac{s_r^d}{b_{ro}^d} + W_2 \sum_{r=1}^{s_2} \frac{s_r^u}{b_{ro}^u}\right)}. \tag{5}$$

Subject to:

$$\begin{aligned} W_1 &= \frac{s w_1}{w_1 + w_2}. \\ W_2 &= \frac{s w_2}{w_2 + w_1}. \\ & (w_1 \geq 0, w_2 \geq 0). \end{aligned}$$

Consequently, if $\rho^* < 1$, the country is inefficient so the excesses in inputs and undesirable outputs must be removed, and the shortfalls in desirable outputs must be increased. A country reaches efficiency when $\rho^* = 1$.

4. Results

Based on the data in Section 3.2, the study utilizes an undesirable outputs model in DEA to analyze inputs and desirable and undesirable variables that relate to EC.

4.1. Data Analysis

Tables A1 and A2 indicate the summarized statistics of input/output factors of 42 countries. In 2017, the values of population, EC, CO_2, CH_4, N_2O, and GDP attained a maximum of 1,386,395,000, 5683.42, 3779.929, 0.2102, 0.0394, and 19,390,604, respectively. The minimum values of population, EC,

CO_2, and GDP are 2,652,340, 18.051, 12.0054, and 29,549.44, in 2008, 2009, 2009, and 2008, respectively. CH_4 and N_2O have minimum values of 0.0007, 0.0001, respectively, within 2008–2010.

DEA is sensitive to outliers so that the data are tested for measurement errors. The tested results indicate the presence and significance of variables. The outlier detection in the data is checked by using the SPSS software. Table A3 denotes that all cases are valid. Electricity consumption, GDP, CO_2, CH_4, and N_2O have a small difference excluding population as shown in Figure A1; however, the populations are important for the electricity consumption, so this factor is still kept to take part in the analysis process.

Moreover, before the data are applied to analysis by models in DEA, they must be checked via Pearson correlation between input variables and output variables to ensure "isotonicity". The values of the correlation coefficient range from −1 to +1. We have a perfect linear relationship between two variables if the correlation coefficient is equal to 1. On the contrary, the variable must be removed and reselected when the correlation coefficient is not positive and significant. As shown in Tables A4 and A5, the Pearson correlations of 42 countries in the research range from 0.303741 to 1; thus, the input and output factors have a standard qualification.

4.2. Efficienct and Inefficient Terms

As per the math in Section 3.3, the countries acquire efficiency when their scores are equal to 1; they are inefficient if their scores are under 1. Table 2 denotes the scores of every country in each term; the scores account for efficient and inefficient terms as well. Belgium, Czech Republic, France, Italy, Poland, Romania, Spain, Sweden, Turkey, Kazakhstan, Russia, Ukraine, Uzbekistan, Canada, Argentina, Brazil, Chile, China, India, Indonesia, Malaysia, South Korea, Thailand, Egypt, South Africa, Iran, Saudi Arabia, and United Arab Emirates are inefficient countries in whole terms because their scores are always lower than 1. Germany achieved efficiency during the period of 2008–2011 and 2013–2014 with its score at 1; however, it proved inefficient in 2012, 2015, 2016, and 2017, as its scores are 0.9062, 0.776, 0.9115, and 0.9861, respectively. The Netherlands attained performance except it remained inefficient in 2015 with a score of 0.9601. Portugal remained efficient from 2008 to 2016, but the growth of modern society led to increased consumption of electricity, which further led to increased CO_2, CH_4, and N_2O emissions in 2017; as a consequence, it remained inefficient in 2017 with a score of 0.9999. Colombia approached efficiency from 2011 to 2013 and excluding inefficient terms from 2014 to 2017, had scores of 0.7522, 0.7069, 0.9692, 0.7257, 0.5765, 0.517, and 0.5291, respectively. Mexico remained inefficient for nine years, as its scores were from 0.2847 to 0.3641, although its score reached 1 in 2015. Japan was efficient during 2009–2011 and inefficient in 2008 and 2012–2017, when its scores were 0.8572, 0.8379, 0.7002, 0.6128, 0.7925, and 0.6942, respectively. Australia achieved efficient performance status during 2010–2015 and in 2017; its scores in 2008, 2009, and 2016 were 0.6228, 0.5712, and 0.8924, respectively. New Zealand remained efficient from 2009 to 2017, excluding 2008, as its score is 0.7689. Algeria achieved efficiency from 2008 to 2010, but remained inefficient during 2011–2017, as its scores are under 1. Besides, five countries, including the United Kingdom, Norway, United States, Nigeria, and Kuwait, were assigned as efficient in the whole term, as their results compute to be 1. Further, these results reveal the ratio among inputs and desirable and undesirable outputs at the balance level.

Table 2. Scores of 42 countries over the period of 2008–2017.

Country	2008	2009	2010	2011	2012	2013	2014	2015	2016	2017
Belgium	0.8017	0.8902	0.8611	0.8847	0.8534	0.856	0.8389	0.8116	0.8396	0.8344
Czech Republic	0.5408	0.6169	0.6054	0.6107	0.5418	0.5202	0.4939	0.5161	0.5019	0.5129
France	0.7824	0.8875	0.8604	0.8677	0.7476	0.7681	0.6953	0.5982	0.6601	0.7055
Germany	1	1	1	1	0.9062	1	1	0.776	0.9115	0.9861
Italy	0.8414	1	0.9036	0.8661	0.7512	0.7741	0.6914	0.5955	0.6733	0.6911
Netherlands	1	1	1	1	1	1	1	0.9601	1	1
Poland	0.3474	0.338	0.3516	0.3518	0.3402	0.3395	0.3338	0.321	0.3163	0.3242
Portugal	1	1	1	1	1	1	1	1	1	0.9999
Romania	0.6265	0.6841	0.569	0.558	0.4867	0.5193	0.4855	0.514	0.4696	0.5932
Spain	0.6551	0.7297	0.6761	0.6463	0.546	0.5723	0.5387	0.4956	0.5447	0.5649
Sweden	0.6287	0.6061	0.6952	0.7498	0.7034	0.7421	0.7481	0.7491	0.8222	0.7904
United Kingdom	1	1	1	1	1	1	1	1	1	1
Norway	1	1	1	1	1	1	1	1	1	1
Turkey	0.3291	0.3074	0.3542	0.3248	0.3417	0.3569	0.3143	0.301	0.3051	0.2696
Kazakhstan	0.3377	0.3816	0.3819	0.3657	0.3616	0.3893	0.3342	0.3383	0.2797	0.2683
Russia	0.1705	0.1386	0.1823	0.235	0.2442	0.2494	0.2008	0.1417	0.14	0.1706
Ukraine	0.1298	0.1378	0.1293	0.1278	0.1296	0.13	0.1211	0.1162	0.1142	0.1204
Uzbekistan	0.1632	0.2047	0.2181	0.2153	0.2231	0.2217	0.2197	0.2585	0.242	0.1755
Canada	0.4752	0.475	0.6068	0.588	0.5354	0.5375	0.5343	0.4995	0.5223	0.5243
United States	1	1	1	1	1	1	1	1	1	1
Argentina	0.2914	0.3075	0.3575	0.3892	0.3976	0.3818	0.3284	0.4023	0.3427	0.4249
Brazil	0.2634	0.3039	0.4215	0.4334	0.3803	0.3565	0.2991	0.2273	0.2448	0.2791
Chile	0.4335	0.5036	0.5549	0.5356	0.5146	0.4716	0.4202	0.4461	0.429	0.4385
Colombia	0.7522	0.7069	0.9692	1	1	1	0.7257	0.5765	0.517	0.5291
Mexico	0.3388	0.2847	0.3641	0.3472	0.3424	0.3637	0.3281	1	0.2868	0.3019
China	0.1257	0.1438	0.1609	0.1758	0.1946	0.2228	0.2274	0.2298	0.2181	0.2168
India	0.0903	0.1057	0.1317	0.1167	0.1113	0.1032	0.0932	0.0944	0.1046	0.1132
Indonesia	0.2483	0.2593	0.3491	0.3387	0.328	0.2885	0.246	0.2482	0.2663	0.2718
Japan	0.8572	1	1	1	1	0.8379	0.7002	0.6128	0.7925	0.6942
Malaysia	0.2642	0.2551	0.2819	0.2769	0.2691	0.25	0.2404	0.2407	0.2218	0.217
South Korea	0.2453	0.2345	0.308	0.2986	0.277	0.2896	0.3287	0.3489	0.3744	0.3975
Thailand	0.1782	0.1946	0.2041	0.2005	0.1966	0.1956	0.1785	0.1881	0.188	0.1927
Australia	0.6228	0.5712	1	1	1	1	1	1	0.8924	1
New Zealand	0.7689	1	1	1	1	1	1	1	1	1
Algeria	1	1	1	0.7032	0.5665	0.4903	0.381	0.3433	0.3091	0.289
Egypt	0.1503	0.1708	0.1745	0.1497	0.1612	0.1602	0.1519	0.1705	0.1639	0.1156
Nigeria	1	1	1	1	1	1	1	1	1	1
South Africa	0.1249	0.1489	0.171	0.1704	0.1684	0.1515	0.1412	0.145	0.1369	0.1475
Iran	0.1824	0.1943	0.2128	0.2248	0.2245	0.169	0.143	0.1465	0.1519	0.1421
Kuwait	1	1	1	1	1	1	1	1	1	1
Saudi Arabia	0.2933	0.261	0.3057	0.3363	0.3482	0.3261	0.3033	0.2857	0.2778	0.2721
United Arab Emirates	0.6857	0.6151	0.6582	0.6598	0.6818	0.6636	0.6117	0.6075	0.5588	0.5568

The above analysis results point out the efficient and inefficient terms in every year, where there are 12 efficient countries and 30 inefficient countries during the period from 2009 to 2011; from 2012 to 2013, there are 11 efficient countries and 31 inefficient countries; 2014 has 10 efficient countries, and 32 inefficient countries; 2008 and 2015 have nine efficient countries and 33 inefficient countries; the period of 2016–2017 has eight efficient countries and 34 inefficient countries. Thus, the quantity of inefficient countries is more than that of efficient countries. The empirical results indicate that United Kingdom, Norway, United States, Nigeria, and Kuwait always approach the efficiency without fluctuation.

4.3. Ranking Countries

Based on the scores shown in Table 2, this study gives in Table 3 the position of each of country in every year.

Table 3. Raking countries during the period from 2008 to 2017.

Country	2008	2009	2010	2011	2012	2013	2014	2015	2016	2017
Belgium	12	13	15	13	13	12	11	11	11	11
Czech Republic	21	18	21	20	20	20	20	19	20	21
France	13	14	16	14	15	15	15	16	15	13
Germany	1	1	1	1	12	1	1	12	9	10
Italy	11	1	14	15	14	14	16	17	14	15
Netherlands	1	1	1	1	1	1	1	10	1	1
Poland	24	25	29	27	30	29	25	28	25	25
Portugal	1	1	1	1	1	1	1	1	1	9
Romania	19	17	22	22	23	21	21	20	21	16
Spain	17	15	18	19	19	18	18	22	17	17
Sweden	18	20	17	16	16	16	12	13	12	12
United Kingdom	1	1	1	1	1	1	1	1	1	1
Norway	1	1	1	1	1	1	1	1	1	1
Turkey	27	27	28	31	29	27	29	29	27	31
Kazakhstan	26	24	25	26	26	24	24	27	29	32
Russia	36	40	37	34	34	34	36	40	39	37
Ukraine	39	41	42	41	41	41	41	41	41	40
Uzbekistan	37	34	34	36	36	36	35	31	33	36
Canada	22	23	20	21	21	19	19	21	18	20
United States	1	1	1	1	1	1	1	1	1	1
Argentina	29	26	27	25	24	25	27	24	24	23
Brazil	31	28	24	24	25	28	31	35	32	28
Chile	23	22	23	23	22	23	22	23	22	22
Colombia	15	16	13	1	1	1	13	18	19	19
Mexico	25	29	26	28	28	26	28	1	28	26
China	40	39	40	38	38	35	34	34	35	34
India	42	42	41	42	42	42	42	42	42	42
Indonesia	32	31	30	29	31	32	32	32	31	30
Japan	10	1	1	1	1	13	14	14	13	14
Malaysia	30	32	33	33	33	33	33	33	34	33
South Korea	33	33	31	32	32	31	26	25	23	24
Thailand	35	35	36	37	37	37	37	36	36	35
Australia	20	21	1	1	1	1	1	1	10	1
New Zealand	14	1	1	1	1	1	1	1	1	1
Algeria	1	1	1	17	18	22	23	26	26	27
Egypt	38	37	38	40	40	39	38	37	37	41
Nigeria	1	1	1	1	1	1	1	1	1	1
South Africa	41	38	39	39	39	40	40	39	40	38
Iran	34	36	35	35	35	38	39	38	38	39
Kuwait	1	1	1	1	1	1	1	1	1	1
Saudi Arabia	28	30	32	30	27	30	30	30	30	29
United Arab Emirates	16	19	19	18	17	17	17	15	16	18

As shown in Table 3, five countries including United Kingdom, Norway, United States, Nigeria, and Kuwait are always at the first position for the whole term. Germany with the first ranking is in 2008, 2009, 2010, 2011, 2013, and 2014. Italy only obtains the first ranking in 2009. The Netherlands is mostly in the first position except for 2015. Portugal obtains first ranking from 2008 to 2016 and it is down to the ninth. Colombia gets the first ranking for three years as 2011, 2012, 2013. Japan attained first position during the period from 2009–2014. Australia is in the first ranking from 2010 to 2015, and in 2017. New Zealand reaches the first position except for 2008. Algeria approaches the first ranking in three years from 2008 to 2011. The remaining terms of Germany, Italy, Netherlands, Portugal, Colombia, Japan, Australia, New Zealand, Algeria, are ranked from 9 to 27. Belgium, Czech Republic, France, Poland, Romania, Spain, Sweden, Turkey, Kazakhstan, Russia, Ukraine, Uzbekistan, Canada, Argentina, Brazil, Chile, Mexico, China, India, Indonesia, Malaysia, South Korea, Thailand, Egypt, South Africa, Iran, Saudi Arabia, and United Arab Emirates stay at the low position without reaching the first position during whole term. Especially, India rank at the bottom position consecutively during the period of 2008–2017 except for 2010 where it raised one level with a ranking as forty-first. Ukraine has the last ranking in 2010.

The above description specifies the ranking of an effect level in electrical energy sources. Increased population, simultaneously industrialization, and modernization all represent an important force that has an impact on accreting emissions. Therefore, the number of efficient countries with first contemporaneous ranking are reduced, from 2009 to 2017 down from 12 to eight countries. Furthermore, many countries such as the Czech Republic, Turkey, Kazakhstan, i.e., have yet to reach first position and thus face a downward trend. On the contrary, the United Kingdom, Norway, United States, Nigeria, and Kuwait maintain a sustainable economy and always stand at the highest ranking.

4.4. Discussion

The empirical results given in Section 4.2 point out the relationship between input and output factors of 42 countries during 2008–2017 when using electricity and reveal their positions in every year as well. The interplay pathway among selected inputs into selected desirable and undesirable outputs in the context of human growth activities in every country is explored based on Table 2. Most countries exhibit a fluctuation, according to each term; however, the United Kingdom, Norway, United States, Nigeria, and Kuwait always approach high scores as 1 and keep a stable position. They obtain an excellent interplay under all the circumstances.

On the other hand, other countries demonstrate a variation in each period. Portugal, The Netherlands, and New Zealand achieve good relations with scores of 1 over nine years, while Portugal kept in balance from 2008 to 2016 and displayed a downward trend in 2017 at 0.9999. The Netherlands dropped in 2015, as its score is only at 0.9601, and the primary score in 2008 is only 0.7689, but its efforts to improve the interplay with upward mobility helped it reach to the high point in the next terms. Italy and Japan achieved a forward movement to obtain a maximum score in 2009; however, both they could not maintain a good relationship, which is down by the end. Algeria and Germany started with a brilliant mark with a maximum value in primal years; Algeria kept it in three years, consecutively, and dropped in the remaining years from 2011 to 2017; Germany has more flourish with a maximum score in six years and an upward trend in the final term from 0.776 to 0.9861. Australia, Colombia, and Mexico fell in 2007, though they pushed up their scores in the next terms; particularly, Australia increased from 2009 to 2010 and held a stable score with a high position over six years consecutively; Colombia augmented in the first terms and decreased in the final terms; as its maximum score of 1 is for only three years from 2011 to 2013, Mexico has a sharp variation from 0.3281 to 1 within one year and then dropped deeply to 0.2868 in the next year. Consequently, these countries fluctuated over time; however, they still display a good interplay during some terms.

Besides, the 27 remaining countries have seen variations every year, thus failing reach to an excellent relationship. Their scores are usually lower than the standard value. Eight countries, i.e., Canada, Czech Republic, Romania, Sweden, Spain, United Arab Emirates, France, and Belgium, are at an average level with most of their values being under 0.5. Nineteen countries, i.e., India, South Africa, China, Ukraine, Egypt, Uzbekistan, Russia, Thailand, Iran, South Korea, Indonesia, Brazil, Malaysia, Argentina, Saudi Arabia, Turkey, Kazakhstan, Poland, and Chile, are seriously affected by emissions, as their valuations are all under 0.5.

As a consequence, the economic development is accreting into producing emissions which are harmful for the environment. According to Chung's directional distance function [15], the performance in this case is refined by increasing the good output while simultaneously reducing the bad outputs. In the study, CO_2, CH_4, and N_2O must decline, but at the same time the GDP still must increase. In addition, the electricity consumption can be reduced when the electricity usage should the saved and replace high-capacity equipment with low-capacity equipment in order to diminish energy consumption. That way, emissions can dwindle to avoid a contaminated environment and climate change, the effect of electricity consumption on climate change was tested by Philli-Sihvola [44]; further, with the inefficient terms, the performance among inputs and desirable and undesirable outputs can be improved.

5. Conclusions

Electricity provides humans with light and operation of machines. Then, if a population is at a high level, the consumption of electricity will increase. As a result, economic growth will be enhanced by displayed in the GDP index; however, electricity production and use brings disadvantages of emitting undesirable factors (CO_2, CH_4, N_2O). Therefore, the study proposes an undesirable outputs model to measure the performance of the elements that relate to the EC process.

For the characteristics of dealing with fixed bad and good outputs, an undesirable outputs model is used help the study formulate scores. The empirical values demonstrate interplay among variables, ranking, and variable pathways of every country in every year. Forty-two countries are defined as efficiency or inefficient after applying an undesirable outputs model to analyzing their performance. The analysis results denote that the United Kingdom, Norway, United States, Nigeria, and Kuwait show stable efficiency and retain a good relationship for the whole term; other countries have changed consecutively every time.

For the 42 countries we not only know about the interplay among inputs, desirable and undesirable outputs but can also understand the quantitative analysis that affect level of emissions. Based on the principle of undesirable outputs model, desirable outputs i.e., GDP should be increased; undesirable outputs including CO_2, CH_4, and N_2O, and inputs, i.e., electricity consumption at the inefficient terms will be reduced, by the way the efficiency will be improved. In addition, they find t a direction to restore balance to their ecosystems.

In general, the study summarizes the basic data of EC and specifies a relationship between EC and related factors; however, limitations remain. First, the inputs and outputs of all countries are not listed, so that the future research should expand to add more countries. Second, the interplay will become deeper when calculations include enough factors. Further study should investigate this in order to obtain more inputs, i.e., capital, assets, and output variables, i.e., revenue. Third, the study only needs the efficiency in the past term through the undesirable outputs model, so further studies could utilize more models to predict the future terms. Fourth, the future direction will use the Spearman correlation coefficient to have a statistical measure of a relationships between paired data.

Author Contributions: C.-N.W. guided the analysis method, and the research direction, found the solutions, and edited the content; Q.C.L. designed research framework, analyzed the empirical result and wrote; T.K.L.N. collected and analyzed the data. All authors contributed in issuing the final result.

Funding: This research was partly supported by MOST107-2622-E-992-012-CC3 from the Ministry of Sciences and Technology in Taiwan.

Acknowledgments: The authors appreciate the support from National Kaohsiung University of Science and Technology, Ministry of Sciences and Technology in Taiwan.

Conflicts of Interest: The authors declare no conflict of interest

Appendix A

Table A1. Statistics of the 42 countries over the period of 2008–2011.

Years	Population	EC (TWh)	CO_2 (Mtons)	CH_4 (Mtons)	N_2O (Mtons)	GDP (Million in USD)
2008	1,324,655,000	3907.229	2598.6199	0.1445	0.0271	14,718,582
	2,652,340	18.517	12.3153	0.0007	0.0001	29,549.44
	118,126,627.26	373.5982	248.4727	0.0138	0.0026	1,336,946.326
	263,672,064.05	728.7291	484.6632	0.027	0.0051	2,411,815.529
2009	1,331,260,000	3724.658	2477.1955	0.1378	0.0258	14,418,739
	2,818,939	18.051	12.0054	0.0007	0.0001	33,689.22
	119,287,116.55	371.3682	246.9896	0.0137	0.0026	1,268,097.807
	266,092,473.27	728.332	484.3991	0.0269	0.0051	2,386,203.268
2010	1,337,705,000	3894.367	2598.6199	0.1445	0.0271	14,964,372
	2,998,083	20.876	12.3153	0.0007	0.0001	39,332.77
	120,434,614.83	397.1979	248.4727	0.0138	0.0026	1,393,321.39
	268,462,393.21	786.178	484.6632	0.027	0.0051	2,506,496.099
2011	1,344,130,000	4051.605	2694.6415	0.1499	0.0281	15,517,926
	3,191,051	23.679	15.7484	0.0009	0.0002	45,915.19
	121,536,746.24	409.9938	272.6786	0.0152	0.0029	1,549,594.858
	270,791,913.9	827.6549	550.4567	0.0306	0.0057	2,671,950.305

Table A2. Statistics of the 42 countries over the period of 2012–2017.

Years	Population	EC (TWh)	CO_2 (Mtons)	CH_4 (Mtons)	N_2O (Mtons)	GDP (Million in USD)
2012	1,350,695,000	4326.079	2877.188	0.16	0.03	16,155,255
	3,395,556	25.399	16.8924	0.0009	0.0002	51,821.57
	122,670,658.43	419.9453	279.1316	0.0155	0.0029	1,583,732.445
	273,099,786.91	851.7249	566.4128	0.0315	0.0059	2,799,627.018
2013	1,357,380,000	4717.568	3137.5601	0.1745	0.0327	16,691,517
	3,598,385	23.689	15.7551	0.0009	0.0002	57,690.45
	123,810,535.81	432.4245	287.745	0.016	0.003	1,622,822.758
	275,389,972.61	899.751	598.4121	0.0333	0.0062	2,899,088.887
2014	1,364,270,000	4938.623	3284.5794	0.183	0.0343	17,427,609
	3,782,450	24.625	16.3776	0.001	0.0002	63,067.08
	124,952,217.6	441.3991	292.8808	0.0163	0.0031	1,662,353.067
	277,690,414.5	927.2612	616.5223	0.0343	0.0064	3,046,071.083
2015	1,371,220,000	5103.889	3301.0023	0.1836	0.0344	18,120,714
	3,935,794	25.268	17.848	0.001	0.0002	66,903.8
	126,084,336.38	448.5672	292.5948	0.0163	0.0031	1,572,917.053
	279,993,795.16	946.811	619.273	0.0344	0.0065	3,157,505.409
2016	1,378,665,000	5366.78	3471.2873	0.1931	0.0362	1,862,4475
	4,052,584	24.5605	16.2416	0.0009	0.0002	67,067.57
	127,216,593.57	459.7215	307.2145	0.0171	0.0031	1,596,048.935
	282,345,525.38	977.6329	635.8452	0.0354	0.0066	3,243,650.311
2017	1,386,395,000	5683.42	3779.929	0.2102	0.0394	19,390,604
	4,136,528	24.4774	16.2794	0.0009	0.0002	48,717.69
	128,333,654.5	471.4782	313.5707	0.0174	0.0033	1,692,506.563
	284,721,026.7	1010.7354	672.2199	0.0374	0.007	3,412,183.899

Table A3. Case Processing Summary.

Factors	Cases					
	Valid		Missing		Total	
	N	Percent	N	Percent	N	Percent
(I) Population	417	100.00%	0	0.00%	42	100.00%
(I) Electricity consumption (TWh)	417	100.00%	0	0.00%	42	100.00%
(O) GDP (million USD)	417	100.00%	0	0.00%	42	100.00%
(Obad) CO_2 (Mtons)	417	100.00%	0	0.00%	42	100.00%
(Obad) CH_4 (Mtons)	417	100.00%	0	0.00%	42	100.00%
(Obad) N_2O (Mtons)	417	100.00%	0	0.00%	42	100.00%

Table A4. Person's correlation over the period of 2008–2012.

Indicators	Year	Population	EC (TWh)	CO_2 (Mtons)	CH_4 (Mtons)	N_2O (Mtons)	GDP (Million USD)
Population	2008	1	0.580126	0.580126	0.580341	0.579862	0.303741
EC (TWh)		0.580126	1	1	0.999999	0.999986	0.901139
CO_2 (Mtons)		0.580126	1	1	0.999999	0.999986	0.901139
CH_4 (Mtons)		0.580341	0.999999	0.999999	1	0.999985	0.901071
N_2O (Mtons)		0.579862	0.999986	0.999986	0.999985	1	0.901355
GDP (million USD)		0.303741	0.901139	0.901139	0.901071	0.901355	1
Population	2009	1	0.616812	0.616812	0.616607	0.616654	0.33962
EC (TWh)		0.616812	1	1	0.999999	0.999985	0.894023
CO_2 (Mtons)		0.616812	1	1	0.999999	0.999985	0.894023
CH_4 (Mtons)		0.616607	0.999999	0.999999	1	0.999984	0.894111
N_2O (Mtons)		0.616654	0.999985	0.999985	0.999984	1	0.893678
GDP (million USD)		0.33962	0.894023	0.894023	0.894111	0.893678	1
Population	2010	1	0.633979	0.576015	0.576229	0.575756	0.380058
EC (TWh)		0.633979	1	0.994505	0.994535	0.994407	0.900126
CO_2 (Mtons)		0.576015	0.994505	1	0.999999	0.999986	0.934326
CH_4 (Mtons)		0.576229	0.994535	0.999999	1	0.999985	0.93427
N_2O (Mtons)		0.575756	0.994407	0.999986	0.999985	1	0.934449
GDP (million USD)		0.380058	0.900126	0.934326	0.93427	0.934449	1
Population	2011	1	0.664182	0.664182	0.664216	0.664145	0.419236
EC (TWh)		0.664182	1	1	1	0.999989	0.898789
CO_2 (Mtons)		0.664182	1	1	1	0.999989	0.898789
CH_4 (Mtons)		0.664216	1	1	1	0.999989	0.898727
N_2O (Mtons)		0.664145	0.999989	0.999989	0.999989	1	0.899497
GDP (million USD)		0.419236	0.898789	0.898789	0.898727	0.899497	1
Population	2012	1	0.680572	0.680718	0.680613	0.680276	0.440727
EC (TWh)		0.680572	1	0.999997	0.999997	0.999983	0.902231
CO_2 (Mtons)		0.680718	0.999997	1	1	0.999989	0.902234
CH_4 (Mtons)		0.680613	0.999997	1	1	0.999988	0.902301
N_2O (Mtons)		0.680276	0.999983	0.999989	0.999988	1	0.90262
GDP (million USD)		0.440727	0.902231	0.902234	0.902301	0.90262	1

Table A5. Person's correlation over the period of 2013–2017.

Indicators	Year	Population	EC (TWh)	CO_2 (Mtons)	CH_4 (Mtons)	N_2O (Mtons)	GDP (Million USD)
Population	2013	1	0.695153	0.696013	0.695992	0.695301	0.464744
EC (TWh)		0.695153	1	0.999998	0.999998	0.999988	0.907294
CO_2 (Mtons)		0.696013	0.999998	1	1	0.999988	0.907197
CH_4 (Mtons)		0.695992	0.999998	1	1	0.999987	0.907255
N_2O (Mtons)		0.695301	0.999988	0.999988	0.999987	1	0.90767
GDP (million USD)		0.464744	0.907294	0.907197	0.907255	0.90767	1
Population	2014	1	0.706217	0.705663	0.705031	0.705695	0.480833
EC (TWh)		0.706217	1	0.999991	0.999952	0.99998	0.907172
CO_2 (Mtons)		0.705663	0.999991	1	0.999961	0.999989	0.907172
CH_4 (Mtons)		0.705031	0.999952	0.999961	1	0.999947	0.907122
N_2O (Mtons)		0.705695	0.99998	0.999989	0.999947	1	0.906706
GDP (million USD)		0.480833	0.907172	0.907172	0.907122	0.906706	1
Population	2015	1	0.714941	0.708444	0.708326	0.70894	0.491955
EC (TWh)		0.714941	1	0.999494	0.999493	0.999839	0.908562
CO_2 (Mtons)		0.708444	0.999494	1	1	0.999616	0.913374
CH_4 (Mtons)		0.708326	0.999493	1	1	0.999616	0.91337
N_2O (Mtons)		0.70894	0.999839	0.999616	0.999616	1	0.914613
GDP (million USD)		0.491955	0.908562	0.913374	0.91337	0.914613	1
Population	2016	1	0.724975	0.718889	0.71891	0.720908	0.490153
EC (TWh)		0.724975	1	0.998089	0.9981	0.999912	0.8947
CO_2 (Mtons)		0.718889	0.998089	1	1	0.998155	0.897223
CH_4 (Mtons)		0.71891	0.9981	1	1	0.998165	0.897175
N_2O (Mtons)		0.720908	0.999912	0.998155	0.998165	1	0.898646
GDP (million USD)		0.490153	0.8947	0.897223	0.897175	0.898646	1
Population	2017	1	0.73631	0.73631	0.736469	0.73605	0.508286
EC (TWh)		0.73631	1	1	1	0.999992	0.891115
CO_2 (Mtons)		0.73631	1	1	1	0.999992	0.891115
CH_4 (Mtons)		0.736469	1	1	1	0.999992	0.89107
N_2O (Mtons)		0.73605	0.999992	0.999992	0.999992	1	0.891228
GDP (million USD)		0.508286	0.891115	0.891115	0.89107	0.891228	1

Appendix B

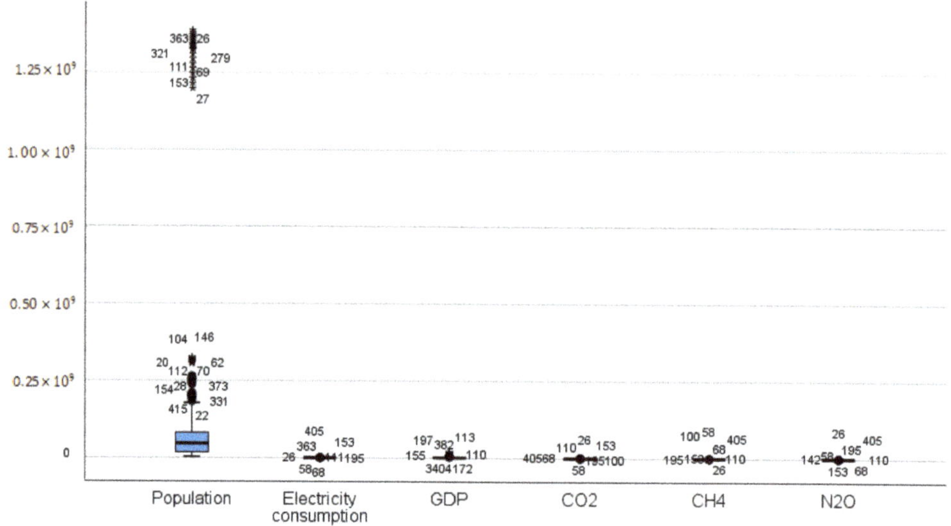

Figure A1. Boxplot of inputs and outputs.

References

1. Richard, F.H.; Jonathan, G.K. Electricity Consumption and Economic Growth: A New Relationship with Significant Consequences. *Electr. J.* **2015**, *28*, 72–84. [CrossRef]
2. Tewathia, N. Determinant of the Household Electricity Consumption: A Case Study of Delhi. *Int. J. Energy Econ. Policy* **2014**, *4*, 337–348.
3. Hameed, L.; Khan, A.A. Population Growth and Increase in Domestic Electricity Consumption in Pakistan: A Case Study of Bahawalpur City. *J. Soc. Sci.* **2016**, *2*, 27–33.
4. Lu, W.-C. Electricity Consumption and Economic Growth: Evidence from 17 Taiwanese industries. *Sustainability* **2016**, *9*, 50. [CrossRef]
5. Enu, P.; Havi, E.D.K. Influence of Electricity Consumption on Economic Growth in Ghana. *Int. J. Econ. Commer. Manag.* **2014**, *II*, 1–20.
6. Altinay, G.; Karagol, E. Electricity Consumption and Economic Growth: Evidence from Turkey. *Energy Econ.* **2005**, *27*, 849–856. [CrossRef]
7. Ologun, O.O.; Wara, S.T. Carbon Footprint Evaaluation and Reduction as a Climate Change Mitigation Tool—Case Study of Federal University of Agriculture Abeokuta, Ogun, State, Nigeria. *Int. J. Renew. Energy Res.* **2014**, *4*, 176–181.
8. To, W.M.; Peter, K.C.L. GHG emissions from electricity consumption: A case study of Hong Kong from 2002 to 2015 and trends to 2030. *J. Clean. Prod.* **2017**, *165*, 589–598. [CrossRef]
9. Hooi, H.L.; Russell, S. CO_2 emissions, electricity consumption and output in ASEAN. *Appl. Energy* **2010**, *87*, 1858–1864. [CrossRef]
10. Mohammed Redha Qader. Electricity Consumption and GHG Emissions in GCC Countries. *Energies* **2009**, *2*, 1201–1213. [CrossRef]
11. Albiman, M.M.; Suleiman, N.N.; Baka, M.O. The Relationship between Energy Consumption, CO_2 Emissions and Economic Growth in Tanzania. *Int. J. Energy Sect. Manag.* **2015**, *9*, 361–375. [CrossRef]
12. Battery and Energy Technologies. Available online: https://www.mpoweruk.com (accessed on 10 August 2018).
13. Seyed Esmaeili, F.; Rostamy-Malkhalifed, M. Data Envelopment Analysis with Fixed Inputs, Undesirable Outputs and Negative Data. *J. Data Envel. Anal. Decis. Sci.* **2017**, *2017*, dea–00140. [CrossRef]

14. Lawrence, M.S.; Joe, Z. Modeling Undesirable Factors in Efficiency Evaluation. *Eur. J. Oper. Res.* **2002**, *142*, 16–20. [CrossRef]
15. Chung, Y.H.; Färe, R.; Grosskopf, S. Productivity and Undesirable Outputs: A Directional Distance Function Approach. *J. Environ. Manag.* **1997**, *51*, 229–240. [CrossRef]
16. Chen, W.-J.; He, G. Electricity Consumption and its Impact Factors: Based on the Nonparametric Model. *Syst. Eng. Theory Pract.* **2009**, *29*, 92–97. [CrossRef]
17. Gajowniczek, K.; Ząbkowski, T. Short term electricity forecasting based on user behavior using individual smart meter data. *Intell. Fuzzy Syst.* **2015**, *30*, 223–234. [CrossRef]
18. Gajowniczek, K.; Ząbkowski, T. Two-stage electricity demand modelling using machine learning algorithms. *Energies* **2017**, *10*, 1547. [CrossRef]
19. Singh, S.; Yassine, A. Big data mining of energy time series for behavioral analytics and energy consumption forecasting. *Energies* **2018**, *11*, 452. [CrossRef]
20. Paul, E.H.; Tom, S.C.; Robert, G.H. Life cycle greenhouse gas emissions from electricity generation: A comparative analysis of Australian energy source. *Energies* **2012**, *5*, 872–897. [CrossRef]
21. Hanqin, T.; Guangsheng, C.; Chaoqun, L. Global Methane and Nitrous Oxide Emissions from Terrestrial Ecosystems due to Multiple Environmental Changes. *Ecosyst. Health Sustain.* **2015**, *1*, 1–20. [CrossRef]
22. Jiang, L.; Ou, X.; Ma, L.; Li, Z.; Ni, W. Life-cycle GHG Emission Factors of Final Energy in China. *Energy Procedia* **2013**, *37*, 2848–2855. [CrossRef]
23. Timo, A.S.; Olli, V.; Laura, S.; Matti, K. Greenhouse Gas Emissions of Hydropower in the Mekong River Basin. *Environ. Res. Lett.* **2018**, *13*, 034030. [CrossRef]
24. Yasutoshi, S.; Satoshi, D.; Kanako, T. The CO_2 Emission Factor of Water in Japan. *Water* **2012**, *4*, 759–769. [CrossRef]
25. Xiaochun, Z.; Nathan, P.M.; Ken, C. Key Factors for Assessing Climate Benefits of Natural Gas Versus Coal Electricity Generation. *Environ. Res. Lett.* **2014**, *9*, 114022. [CrossRef]
26. Chang-Sang, C.; Jae-Hwan, S.; Ki-Kyo, L.; Tae-Mi, Y. Development of Methane and Nitrous Oxide Emission Factors for the Biomass Fired Circulating Fluidized Bed Combustion Power Plant. *Sci. World J.* **2012**. [CrossRef]
27. Yuxuan, W.; Tianye, S. Life Cycle Assessment of CO_2 Emissions from Wind Power Plants: Methodology and Case Studies. *Renew. Energy* **2012**, *43*, 30–36. [CrossRef]
28. Ma, C.M.; Ge, Q.S. Method for Calculating CO_2 Emissions from the Power Sector at the Provincial Level in China. *Adv. Clim. Chang. Res.* **2014**, *5*, 92–99. [CrossRef]
29. Greg, S.; Ines, A.; Constantine, S. Assessing the Evolution of Power Sector Carbon Intensity in the United States. *Environ. Res. Lett.* **2018**, *13*. [CrossRef]
30. Tone, K. *Dealing with Undesirable Outputs in DEA: A Slacks-Based Measure (SBM) Approach*; National Graduate Institute for Policy Studies: Tokyo, Japan, 2003.
31. Amiteimoori, A.; Toloie-Eshlaghi, A.; Homayoonfar, M. Efficiency Measurement in Two-Stage Network Structures Considering Undesirable Outputs. *Int. J. Ind. Math.* **2014**, *6*, 65–72.
32. Scheel, H. Undesirable Outputs in Efficiency Valuations. *Eur. J. Oper. Res.* **2001**, *132*, 400–410. [CrossRef]
33. Gongbing, B.; Liang, L.; Jie, W. Radial and non-radial DEA Models Undesirable Outputs: An Application to OECD Countries. *Int. J. Sustain. Soc.* **2010**, 2. [CrossRef]
34. Tone, K.; Tsutsui, M. Applying an Efficiency Measure of Desirable and Undesirable Outputs in DEA to U.S. Electric Utilities. *J. Cent. Cathedra* **2001**, *4*, 236–249. [CrossRef]
35. Wen-Hsien, T.; Hsin-Li, L.; Chid-Hao, Y.; Chung-Chen, H. Input-Output Analysis for Sustainability by Using DEA Method: A Comparison Study between European and Asian Countries. *Sustainability* **2016**, *8*, 1230. [CrossRef]
36. You, S.; Yan, H. A New Approach in Modelling Undesirable Output in DEA Model. *J. Oper. Res. Soc.* **2001**, *62*, 2146–2156. [CrossRef]
37. Flavia, D.C.C.; Daisy, A.D.N.R.; Roberta, T.R. Energy Efficiency Analysis of BRICS Countries: A Study Using Data Envelopment Analysis. *Gest. Prod.* **2016**, *23*, 192–203. [CrossRef]
38. Electricity Consumption. Available online: https://yearbook.enerdata.net (accessed on 8 July 2018).
39. Population, GDP. Available online: https://data.worldbank.org/indicator (accessed on 8 July 2018).
40. CO_2, CH_4, N_2O. Available online: https://www.epa.gov/sites/production/files/2015-08/sgec_tool_v3_2.xls (accessed on 8 July 2018).

41. Tone, K. A slack-based measure of super-efficiency in data envelopment analysis. *Eur. J. Oper. Res.* **2002**, *143*, 34–41. [CrossRef]
42. Tone, K. A slacks-based measure of efficiency in data envelopment analysis. *Eur. J. Oper. Res.* **2001**, *130*, 498–509. [CrossRef]
43. Charnes, A.; Cooper, W.W. Programming with linear fractional functionals. *Nav. Res. Logist. Q.* **1962**, *9*, 181–186. [CrossRef]
44. Pilli-Sihvola, K.; Aatola, P.; Ollikainen, M.; Tuomenvirta, H. Climate change and electricity consumption—Witnessing increasing or decreasing use and costs? *Energy Policy* **2010**, *38*, 2409–2419. [CrossRef]

MDPI
St. Alban-Anlage 66
4052 Basel
Switzerland
Tel. +41 61 683 77 34
Fax +41 61 302 89 18
www.mdpi.com

Energies Editorial Office
E-mail: energies@mdpi.com
www.mdpi.com/journal/energies

www.ingramcontent.com/pod-product-compliance
Lightning Source LLC
LaVergne TN
LVHW070617170726
843515LV00004B/109

* 9 7 8 3 0 3 6 5 0 5 7 2 5 *